The WRNS

The WRNS

A History of the Women's Royal Naval Service

Commandant M.H. Fletcher CBE

Naval Institute Press

ACKNOWLEDGEMENTS

I extend my grateful thanks to everyone who has given me encouragement and practical assistance in the production of this record of the development of the Women's Royal Naval Service. In particular, I would like to mention all those who wrote to me with dates, events, photography and personal memories:

Mrs H. Adams, Mrs B. Ambrose, Mrs M. E. Andrew, Miss H. L. Andrus, Miss E. Archdale, Mrs A. Asher, Miss C. Avent, Miss C. M. Baker, Mrs M. K. L. Baker, Mrs D. Bailey, Mrs V. Bain, Mrs D. E. Baldwin, Mrs M. I. Barnby, Mrs Y. M. Barrett, Mrs M. Beck, Miss H. M. A. Beekmaster, Mrs. D. E. Bennett, Mrs M. Black, Mrs E. N. Bliss, Mrs E. Bowen-Jones, Mrs S. Broster, Mrs M. Brown, Mrs A. Buckley, Miss M. E. Callanan, Miss J. Carpenter, Mrs M. J. Carter, Miss J. Cole, Mrs J. Colenutt, Mrs C. L. Coulthard, Mrs M. Cowans, Mrs P. Crossley, Mrs M. Currie, Mrs J. Curry, Mrs E. Dennis, Mrs J. R. Dilks, Mrs J. Dinwoodie, Mrs G. Dix, Mr J. A. Dodds, Mrs J. M. Donaghy, Mrs L. Dougray, Mrs E. M. Drury, Miss H. Elliston, Miss H. J. F. Emerson, Mrs L. M. Farrell, Mrs J. Firth, Miss F. Flowers, Mrs E. M. Foxon, Mrs K. A. Franklin, Mrs C. Fraser, Mrs J. Gallagher, Mrs H. Gibling, Mrs E. Gilbert, Mrs V. C. Goodall, Mrs S. Graves, Mrs S. R. G. Hague, Mrs M. Hamilton, Mrs S. M. Hamnett, Mrs J. Hatfield, Miss F. Hayes, Mrs P. Hoad, Mrs M. Hodges, Mrs W. E. W. Hogarth, Mrs M. Holden, Mrs E. Holford-Smith, Miss M. Hopkins, Miss M. R. Hopkins, Miss D. Humphreys, Mrs N. M. Hunt, Mrs P. Inverarity, Mrs D. M. Kellett, Mrs R. L. Kennedy, Miss J. Knight, Mrs E. Jacques, Mrs D. John, Mrs P. Lachlan, Miss J. C. Larcombe, Mrs T. A. Leach, Mrs K. Lindsay-MacDougall, Mrs L. M. Lowry, Mrs P. Mack, Mrs R. D. Matthews, Miss J. McCormack, Mrs M. McGuiness, Mrs M. Mears, Mrs B. M. Millott, Mrs J. A. Mitchell, Miss P. Neale, Mrs E. Nelson-Ward, Mrs J. Oxtoby, Mrs C. L. Pearn, Mrs J. Perry, Mrs M. Picco, Mrs G. Pope, Mrs A. Porter, Mrs M. Powell, Mrs I. M. Preston, Mrs J. E. Pritchard, Mrs K. Roberts, Dame Nancy Robertson, Mrs G. Robinson, Mrs W. Robinson, Miss E. H. Scott, Mrs M. Shaw-Reynolds, Mrs J. Shea, Mrs J. Shead, Mrs L. P. Sheridan, Mrs E. M. Singleton, Miss K. Skin, Mrs M. Somers, Miss M. Somerset, Mrs J. M. Stevens, Mrs P. I. Streater, Mrs M. Strickland, Mrs F. E. Sutton, Mrs C. B. C. Vane Percy, Mrs D. Wakem, Mrs K. Wallace, Mrs M. Williams, Mrs M. L. Wilson, Miss A. Winser, Miss E. F. Wood, Mrs P. Weitz.

In addition I am grateful to the Ministry of Defence (Navy) for allowing me to use the official history of the WRNS, as a source document on the Second World War and official photography, Director WRNS for the invaluable information reposing in her historical cabinet, and the Association of Wrens for permission to reprint articles from *The Wren*.

Lastly, to Dr Christopher Dowling of the Imperial War Museum, without whose support and practical advice the book would not have been launched.

CONTENTS

LIST OF ABBREVIATIONS

AA	anti-aircraft
ADP	automatic data processing
ADR	Air Defence Region
ANCXF	Allied Naval Command Expeditionary Force
ARP	air-raid precautions
CO	Commanding Officer
CPO	Chief Petty Officer
CPOOW	Chief Petty Officer of the Watch
EVT	Educational and Vocational Training
HMTE	Her Majesty's Training Establishment
LCI	landing craft, inshore
LCT	landing craft tank
MS	motor-ship
MTB	motor torpedo-boat
NCS	Naval Control Service
NOIC	Naval Officer-in-Charge
OTC	Officer Training Course
PO	Petty Officer
QARNNS	Queen Alexandra's Royal Naval Nursing Service
RFA	Royal Fleet Auxiliary
RNVR	Royal Naval Volunteer Reserve
R/T	radio-telephony
SBA	Sick Berth Attendant
SDO	Signal Distribution Office
SS	steamship
SSAFA	Soldiers', Sailors', and Airmen's Families Association
T/P	tele-printer
USN	United States Navy
V/S	Visual Signaller
WAAF	Women's Auxiliary Air Force
WATU	Western Approaches Tactical Unit
WRAC	Women's Royal Army Corps
WRAF	Women's Royal Air Force
WRANS	Women's Royal Australian Naval Service
WRNZNS	Women's Royal New Zealand Naval Service

Unless otherwise indicated, all photographs come from the collection of the Director WRNS. M.H.F. indicates that photographs come from the Author's private collection.

BUCKINGHAM PALACE

It is now seventy years since the first formation of the Women's Royal Naval Service and, for the last forty years, it has been a permanent and integral part of the Royal Navy. As with any organisation the WRNS has seen many changes and, through these changes, has evolved into a highly professional support arm to the Navy.

This book not only records the major landmarks in that evolution, but complements them with some memories of Wrens showing the strong ties which the Service engenders in any one who wears a Wren's uniform.

I am sure that anyone who has had connections with the Royal Navy will find in these pages references and anecdotes which strike a familiar chord and, for those serving today, examples of the experiences of their predecessors and excellent information for their own careers.

Anne

INTRODUCTION

When I became Director of the Women's Royal Naval Service I inherited a large cupboard containing documents and memorabilia, all of which were gathering dust. Amongst the contents were books which had been written about the Service and it occurred to me that there did not appear to be a published record which set out the factual development of the Service. Fired with enthusiasm I decided to see what I could glean from the material to hand. The cupboard proved to be a treasure trove of information and, subsequently, I let it be known that I had it in mind to try to set down the beginning and continuing history of the Service which had been a central part of most of my working life. An immediate, and favourable, reaction came from members of the Association of Wrens and serving personnel.

Once I had started to compile the chronological data, and word spread of my efforts, letters started to come in from ex-Wrens with details of places and events. It was evident that they were also taking the opportunity to record memories of their own time in the Service. I realised that here was a further fund of information which would not see the light of day unless I incorporated these memories in the book.

Throughout history women have been closely involved in defence of the home, whether this be in the survival of the family base or, on occasion, taking up arms alongside their men. Perception of woman's place in the scheme of things has altered as social fashions changed. There was a long period during which men took their families with them to the war zones; and even in the last century this was still the practice. The harshness of their lives has been recorded in many ways but, for my purposes I would wish to mention that it was quite often the women's responsibility to find their own accommodation, whether in town or in the field, forage for food and fuel, follow the battle lines – often with little military protection – and tend the wounded. All this done with the knowledge that they might also be subjected to a variety of indignities if captured by the enemy.

By the turn of the century women increasingly were left at home when their men went off to do battle, and they had to survive as best they could. Without the wives and camp followers it seems that the wounded had even less chance of survival. The story of Florence Nightingale and her nurses is well known and in 1885 the Naval Nursing Service was formed. By 1902 Army Nurses were approved by Royal Warrant and by 1914 Queen Alexandra's Imperial Military Nursing Service, the Territorial Force Nursing Service, the First Aid Nursing Yeomanry – the FANY, The Women's Convoy Corps, and the Voluntary Aid Detachments – the VAD – were all in being. The first formal steps towards incorporating women into the Armed Forces had been taken.

When the Admiralty took the decision to employ women as a uniformed Service in support of the Royal Navy, Dame Katharine Furse, who had been Commandant of the Women's VAD Organisation, was appointed to be the first Commandant of the Women's Royal Naval Service on its formation in 1917. At that time a substantial number of women were already working in support of the Royal Navy. The first Wrens to appear in uniform were recruited in January 1918 and by the time the Service was disbanded in July 1919 some 7,000 women had worn naval uniform. From early tasks such as clerical and domestic duties the WRNS expanded its skills into servicing anti-submarine equipment, coding, driving, aircraft maintenance, experimental work, signals and the newfangled wireless telegraphy, among others. Wrens served throughout the length and breadth of the country and were established overseas in the Mediterranean.

Although the WRNS was disbanded at the end of the First World War it was decided to form an Association whereby ex-Wrens could keep in touch. This Association proved to be invaluable when in December 1938 the first public announcement appeared regarding the re-formation of the Service. Once again it was thought that only a comparatively small number of women would be required to give shore support to the Royal Navy, and that the type of work would be limited to clerical and domestic duties.

As in the First World War this was to prove a fallacy. As early as 1937 some women had been recruited to train as Cypher Officers and, under the guidance of Mrs Vera Laughton Mathews, who was to be Director of the Service for the duration of the War, Wrens were to work across the whole gamut of naval tasks excluding service at sea in warships, although even here there were to be exceptions. Technological advances since the end of the First World War had been dramatic and Wrens soon found themselves closely involved in learning technical skills far removed from pre-war concepts of what women were capable of undertaking. By 1944 over 74,000 women were in the Service and, apart from service throughout the United Kingdom, they were deployed to the Mediterranean, Middle and Near East, Far East, the Americas and Australia. After 1944 they moved into Europe.

By the end of the War discussions were in hand as to the future of the Women's Services. Although the majority of women were, with the men, demobilised, a significant number were retained to handle the transition from war to peace. Such was the part that women had played in support of the Armed Forces that it was decided there was room for them in peacetime. On 1 February 1949 the WRNS became a permanent and integral part of the Royal Navy. Since that time its rules and organization have become increasingly aligned to that of the Royal Navy. In 1977 the WRNS came under the Naval Discipline Act and, subsequently, conditions of service, pay, promotion and career prospects have all evolved along similar lines to their male peers. There are still differences in that Wrens do not serve at sea on a regular basis and, therefore, there are categories of work which are not open to them. Identical professional training is given to men and women in mutual branches, none more so than in the development of those skills necessary in the computer age, and the interchange of men and women in shore support tasks is now the normal practice. The WRNS has never been more than five per cent of the overall strength of the Royal Navy but it continues to provide that essential factor necessary for the sea-going Navy: a uniformed, disciplined, professional shore support Service.

I have made no attempt to record, in detail, the background to the evolution of the Service but, I hope, that by showing the major dates and events, together with speeches, reports and some memories of Wrens, in the context of the times, the book will complement previous writings and give pleasure to the many women, their families and friends, who have served in or been associated with the Women's Royal Naval Service.

M.H.F. 1989

CHAPTER I
THE FIRST WORLD WAR

The First World War saw the formation of regular Women's Services in support of the Armed Forces and Dame Katharine Furse GBE was invited by the First Lord of the Admiralty to develop a 'Naval organisation of women'. She subsequently had discussions with the Second Sea Lord, the Director of Mobilisation, the Secretary to the Admiralty and an officer of the Royal Naval Air Service, and undertook to develop a scheme.

The choice of name for the Service was very important as other Women's Corps had earned, not particularly attractive, nick-names. The list of alternates was as follows:

Women's Auxiliary Naval Corps	WANKS
Women's Royal Naval Service	WRNS
Women's Naval Service	WNS
Women's Auxiliary Naval Service	WANS
Royal Naval Women's Service	RNWS

There was no doubt about the desired title of Women's Royal Naval Service and its shortened version of 'Wren' and, after consideration, the Admiralty gave its approval including sanctioning the use of the word 'Royal'.

Three women, Dame Katharine, Mrs Tilla Wallace and Miss Edith Crowdy drafted the terms of service, pay, allowances and regulations, and designed a suitable uniform. His Majesty The King approved the formation of the Service and, on 29 November 1917, an Admiralty Office memorandum announced its establishment and recruiting for officers was commenced.

For the duration of its existence the Service was to provide personnel wherever the Admiralty required them to serve. Wrens carried out domestic and clerical duties, officers replaced paymasters, were secretaries to Admirals, coders and decoders; Wrens replaced writers, telephonists, telegraphists, signallers, storekeepers and draughtsmen. They manned listening stations, fitted depth charges and paravanes

A recruitment poster, designed by Joyce Dennys.

in ships, attached floats to torpedo nets, cleaned boilers, washed life-belts and drove cars. They baked, painted, cleaned and ran accommodation hostels and did a myriad of jobs which released men for sea service.

1917–1919

The first public information was issued in late November 1917 through a press release which announced, 'Women for the Navy – New Shore Service to be formed, WOMEN'S ROYAL NAVAL SERVICE, Dame Katharine Furse GBE to be appointed as Director.' Local labour exchanges provided selection boards supplemented by WRNS Recruiting Officers. It was realized from the very first the importance of establishing the Service's reputation and very high standards were set which many potential recruits could not meet.

Dame Katharine Furse GBE

In the early days it was thought that the ratings would be drawn from areas where they could live at home and, therefore, be termed immobile. However applications were received from women who lived far from the port areas and suitable buildings had to be found to accommodate them. Mobile and immobile officers were recruited from the outset, and from these

Divisional Directors were appointed in London, Portsmouth, Chatham, Devonport, Edinburgh and Cardiff. Deputies were established at Immingham and Harwich, and an acting Deputy Divisional Director went to Liverpool. The WRNS Headquarters was in London.

Wrens stationed in the London Division covered a disparate number of duties. At the Royal Naval College, Greenwich 40 ratings gradually took over the cooking for the College; in the War Registry WRNS officers and ratings were closely involved in coding and clerical duties; at the Anti-Aircraft Defence Corps the telephonists continued during all air raids calling up the gun stations, passing through the orders for the gun-fire and barrage; Wrens provided support at two London Air Stations; and at Battersea Experimental Workshops were employed in drawing, tracing and preparing designs for new machines, guns, etc. The first wireless telegraphy course in London provided a new source of professional manpower, as did training of senior Writers to keep naval pay ledgers; and eventually 70 Motor Drivers were working for the Admiralty.

Wrens were established in the Portsmouth area on 22 January 1918 where women were already being employed by naval authorities. In a short time Wrens were working at the various barracks, the Mining School, Paravane Department, Signal, Gunnery and Submarine Schools and in the Commander-in-Chief's office.

Meanwhile, to the east, Chatham was getting organized. In addition to Chatham itself, the Division included Dover, Deal, Sheerness, Broadstairs, Hastings and Folkestone. Catering and clerical tasks were the principal occupations of Wrens in the Royal Naval and Royal Marine barracks, but at the outstations, such as Dover, Wrens were mine-net workers and, at the Experimental Base, they were Anti-Gas respirator workers; they were drivers and despatch riders, scrubbed life-belts, and were porters in the victualling stores.

Down in the south-west substitution of women for men began in the Commander-in-Chief's office in September 1917. They were to be absorbed into the WRNS and a number of WRNS officers eventually replaced naval officers, including watch-keeping in the Confidential Book Room. The Division covered an enormous area from Torquay in the south to Westward-Ho in the north; along the coast of the West Country through Plymouth to Penzance. Many

Above: *An early group photograph – the first Wrens to appear in uniform were enrolled at the RN Depot, Crystal Palace in January 1918.*

Below: *Wren Cooks were one of the earliest categories to be established.*

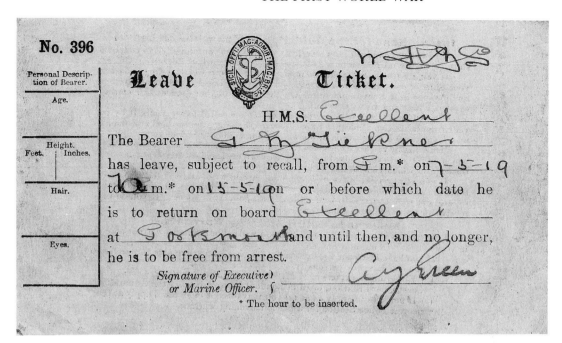

A First World War Leave Ticket. On the reverse were instructions regarding illness and loss of railway ticket.

Ratings serving in the grocery canteen in the Royal Marines Barracks, Chatham.

Wren Porters were used in the victualling stores.

A break from scrubbing lifebelts.

Wrens were attached to Air Stations in the area, both Royal Naval and Royal Air Force, including Seaplanes Scilly, Tresco. They also took over the telephone exchange and message rooms in Plymouth and were at work in offices, the Wardroom, Warrant Officers' Mess, bakery, sausage factory and the stores. In HMS *Apollo*, the depot ship for the 4th Flotilla Destroyers, Wrens worked as sailmakers, turners, fitters and clerks, and further Wrens worked aboard HMS *Indus* and HMS *Powerful* which were moored out in the stream.

In January 1918 the Director sent the WRNS Divisional Director off to Scotland with the following words: 'You are pioneers. Go to your work. Uphold the dignity and honour of the Service.' From the unlikely location of an office in a bedroom, in the North British Hotel in Edinburgh, the work of recruiting and organizing began. Eventually, Wrens were to be found from the Orkneys and Shetlands in the far north to Luce Bay in the south; from Stornoway in the Outer Hebrides to East Fortune in the south-east.

In the Bristol Channel Division demands for women to release men for combatant duties were to increase steadily. As in other areas a large proportion of WRNS personnel were in the clerical category employed across the whole range of office duties, including courts of inquiry, stores ledger work and accounts. Messengers and telephonists were much in evidence working day and night shifts; drivers, in between chauffeuring, washed and maintained their vehicles and, at Air Stations in Pembrokeshire, Wrens were cleaning and repairing aeroplanes.

The Humber was a centre of the anti-submarine offensive and the convoy system. From the spring of 1917 the local naval authorities were employing women to replace and supplement men in every branch of non-combatant work on shore. Initially

Wrens being drilled on Whale Island Parade Ground, 1916.

employed as clerks and telephonists they were quickly absorbed as coders, working in the torpedo sheds, in charge of the stores, as mechanics on naval gyroscopes and searchlights, constructing submarine nets, cleaning mines and priming depth charges. For the duration of the War WRNS officers and ratings worked watches alongside the men, hampered by air raids, in very difficult working conditions, and were the first Wrens to appear on parade beside the men of the Royal Navy when the King and Queen visited the east coast in April 1918.

Further south on the east coast the Naval Base at Gorleston, Great Yarmouth in the Harwich Division was one of the earliest to get started. The majority of the women were employed in mine-net work, with the remainder acting as storekeepers, clerks and telephonists. At Lowestoft women were boiler cleaning on the trawlers and drifters, working on depth charges, sail-making, wirework, storekeeping and in offices. At Ipswich they ran the telephone exchange and at the coastguard station, Harwich, they were clerks and had the sole Wren Driver in the Division. Across the harbour Wrens cooked for the men at the training establishment at Shotley, manned the telephone exchange and were clerks. They were also attached to HMS *Dido* as Anti-Gas respirator workers, and at HMS *Osea* they were employed on domestic and gardening duties.

The last of the original Divisions was started in Liverpool in March 1918 where the appearance of the first two WRNS officers in uniform caused great excitement. An initial reluctance from the civilian women, already working for the naval authorities, to being absorbed into the WRNS was overcome as new Wrens were enrolled. The work in Liverpool was mostly clerical with a few telephone operators and drivers. Two officers went to Holyhead for de-coding duties, to be quietly followed by others. Immobile Wrens were employed at the Royal Naval Air Stations Walney Island, and Llangefni, Anglesey where they worked as cooks, stewards and in general station duties.

Three further Divisions were to be established on the Tyne, in Ireland and the Mediterranean. The Tynemouth Division ran for exactly one year and chiefly employed clerks, telephonists, motor drivers and mine-net workers, in small units, at a number of different offices scattered over a wide area including Newcastle, Tynemouth and Middlesbrough. The Senior Naval Officer was to put on record that 'all

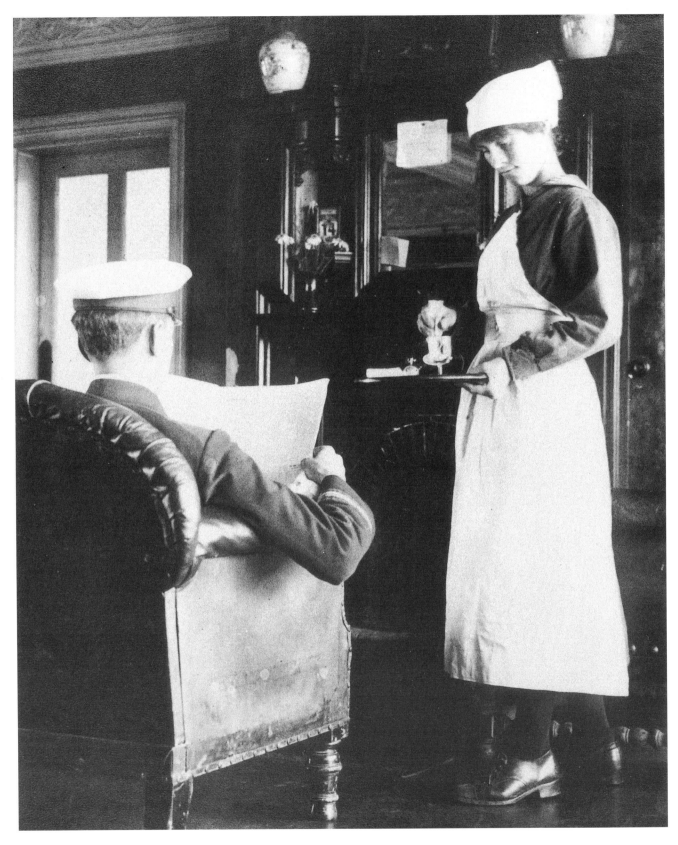

A Wren waitress in an Officer's Mess

ranks and ratings have loyally done their duty and assisted to raise the standard of efficiency in this base.'

In Ireland the Divisional Director established her headquarters at Kingstown, near Dublin, but the bases stretched from Buncrana, Belfast and Larne in the north, through Kingstown and Dublin in the midlands to Queenstown in the south. Larne was an important anti-submarine base and here Wrens were employed as clerks, messengers, storekeepers, mine-net workers and on depth charges and hydrophones. They also helped to unload the ships and, when work was slack, they built a landing stage. They were all immobile. In Buncrana mobile and immobile Wrens were employed as clerks, cooks, and domestic workers; whilst elsewhere in the Division officers were much involved in coding and administrative duties.

The final Division was in the Mediterranean where WRNS officers and ratings were employed in Malta, Gibraltar and Genoa, mostly on cypher duties. A sub-division in Bizerta was almost in being by the armistice, and plans were well ahead to establish further bases in Egypt. Had the war continued there would have been a very large Division in the Mediterranean as there were plans to employ Wrens at Corfu, Taranto, Naples, Syracuse, Marseilles, and possibly Oran and Mudros.

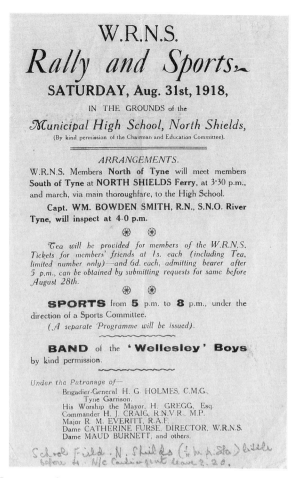

All Service personnel carried a Certificate of Identity.

Organized sports events were a feature of life off duty.

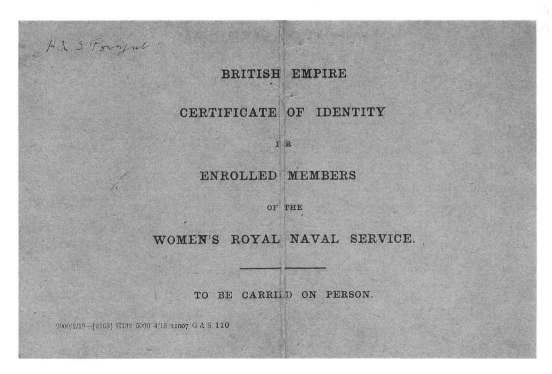

THE FIRST WORLD WAR

The Royal Flying Corps and Royal Naval Air Service were amalgamated on 1 May 1918. At this time some 2,000 Wrens were employed on Air Stations and, although they continued to be looked after by the WRNS administration, they were eventually transferred to the newly formed Women's Royal Air Force.

There was a Procession of Homage of the women's war organizations to Buckingham Palace on 29 June 1918. A WRNS contingent took part in this procession to celebrate the Silver Wedding of King George V and Queen Mary. Afterwards the Second Sea Lord wrote to the Director:

Opposite: *One of the many Wrens who worked on aircraft.*

Dame Katharine Furse leading the Wrens in the Peace March through London, 19 July 1919.

I must express to you my very sincere congratulations on the good appearance, deportment and smartness of the WRNS.

I was very much struck by their general appearance of well-being and contentment.

I hope you will let it be known to all concerned how proud we of the Navy felt of our WRNS.

In October 1918 the mail steamer *Leinster* was torpedoed between England and Ireland. There were three Wrens on board, one of whom was lost at sea. Wren Josephine Carr was the first Wren to lose her life on Active Service.

On 11 November 1918 The King sent a message of thanks to the Royal Navy, the Mercantile Marine and the Fleet Auxiliaries in which he said, 'I wish to express my praise and thankfulness to THE OFFICERS, MEN, AND WOMEN OF THE ROYAL NAVY & MARINES.'

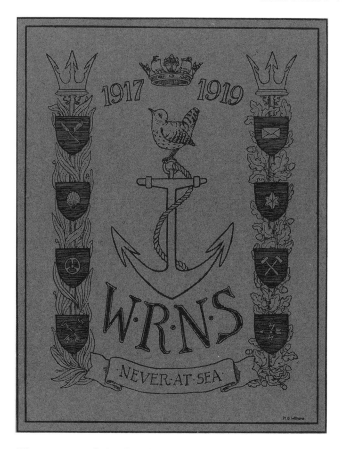

The cover of the booklet The Wrens *showed the badges they had worn, and the slogan 'Never at Sea'.*

A service to commemorate the inauguration of the WRNS was held at St Martin-in-the-Fields on 24 November 1918.

A party of Wrens from the London Hostel joined the public at Buckingham Palace on 29 June 1919 to celebrate the signing of peace. In her report to the Director, Principal M. A. Julius said:

After the band had ceased playing and the Royal Family had withdrawn from the balcony, I received a message from the King's Equerry that H.R.H. Princess Mary wished to speak to the Officer in charge of the Detachment. I therefore went in to the Palace and had the honour of being presented to H.M. The Queen, H.R.H. The Prince of Wales, H.R.H. Princess Mary, H.R.H. Prince Albert and H.R.H. Prince Henry.

I am glad to report that H.M. The Queen seemed pleased with the desire of the London Depot Hostel to express their loyalty and devotion on this auspicious occasion.

Dame Katharine Furse led the WRNS Contingent in the Great Peace Day March through London on 19 July 1919.

For the Wrens themselves one of the proudest moments of the day came when they re-entered Hyde Park and were greeted by the applause of the whole body of Admirals who had headed the Naval Contingent, and who had now fallen out in a group below the Achilles Statue. In that one graceful act was summed up the whole history of the generous treatment accorded by the Navy to the WRNS.

(From *The Wrens*)

The Wrens – Being the Story of their Beginnings and Doings in Various Parts was issued from WRNS Headquarters dated 19 July 1919. Much of the content of that booklet has been used in this chapter to illustrate the sort of work they did and the places in which they were stationed. In her Foreword to *The Wrens* Dame Katharine Furse said:

You know what a wonderful chance the Admiralty gave us and the generous way in which the Navy accepted us. You all realise as well as I do the tremendous honour afforded to us in being allowed to serve with the Navy. It will be one of the greatest regrets of my life that we may not continue in the WRNS, but there is no room for women in the Navy in peacetime. We hope to maintain some form of Reserve and will publish particulars of this in the future, so that all demobilised members may join up again if they like to.

Before stopping, I want to impress two facts. Firstly, the value of preparation, and, secondly, our responsibility. If some of us had not been ready when war broke out, women would not have had such a real chance of helping to win the war.

On 18 September 1919 the Director WRNS received a letter from Their Lordships expressing appreciation for the Wrens' contribution to the war effort.

On the occasion of the general demobilisation of the Women's Royal Naval Service, I am to request that you will communicate to all concerned Their Lordships' high appreciation of the efficient manner in which the Service has been organised and conducted, of the zeal and exemplary conduct which its members have shown in the performance of their duties and of the assistance which they have afforded to the Naval Service generally.

CHAPTER 2
BETWEEN THE WARS

Between the Wars the Association of Wrens was formed and provided a forum for ex-Wrens to keep in touch. In addition to starting a magazine, *The Wren* (first published February 1920), the Association was also concerned with the formation of Service Women's Clubs in London and Edinburgh, and a holiday home in Sussex.

By the mid-Thirties fears of war once again turned thoughts to the part women might play in the Armed Forces if those fears were realized.

In October 1935 a Special Sub-Committee of Imperial Defence considered the question of the creation of a Women's Reserve. In May 1936 the Committee reported: 'Women's Reserve deemed not desirable'.

In September 1937 Dame Katharine Furse wrote to the Board of Admiralty offering the services of the Association of Wrens in the event of an emergency. By 1938 the Admiralty was well ahead with plans for the re-formation of the Service.

On 25 July 1938 an Admiralty Board Minute set out the principles for the employment of women in an emergency; shortly to be followed by the decision to form a Women's Royal Naval Service. An Admiralty letter was sent to Commanders in Chief on 19 August 1938 asking them to report the number of women who would be required to relieve Active Service personnel, the type of work they could do, and how many needed to be uniformed personnel.

A Paper was submitted to the Admiralty Board containing proposals for the re-formation of the WRNS on 22 November 1938, and the first public announcement of these plans appeared in a Ministry of Labour Handbook in December.

CHAPTER 3

THE SECOND WORLD WAR

As the storm clouds gathered and the Armed Forces prepared for the likelihood of confrontation, increased demands were being made on the Navy's trained shore staff. Although the formation of a Women's Reserve had been discounted in 1935, the subject of employment of women in support of the Armed Forces in the event of war gathered momentum. In the case of the Royal Navy the initial outline plan envisaged a need for about 3,000 women who would be mostly clerks, domestics, drivers etc., and in peacetime would be employed as Civil Servants. If war was declared there would be a need to put some of them in uniform, but still on a civilian basis. In the autumn of 1938, however, a paper was circulated within the Admiralty on the formation and organization of 'a Corps to be known as the Women's Royal Naval Service'. At the end of 1938 a handbook on National Service was published which included information about 'Women's Service in the Royal Navy'.

Early in 1939 Mrs Vera Laughton Mathews was appointed Director of the WRNS. With a very small staff she set about the organization of the Service, and by the outbreak of war a nucleus of officers was ready to take the strain. By December 1939 there were over 3,000 WRNS personnel and the strength was to steadily increase until it reached its peak of 74,620 in 1944.

As will be seen in the following pages Wrens were to be employed across a whole range of usual and unusual skills, many of which only exist in wartime. They were to prove themselves capable of coping with dangerous and difficult circumstances, developing considerable endurance, turning their hands to an infinite variety of tasks and earning a reputation for high standards and loyalty to the Royal Navy. Their qualities were tested throughout the United Kingdom and abroad, at sea aboard troopships and, after VE Day, on the Continent and in Scandinavia.

1939

Admiralty proposals were submitted to the Treasury on 11 February 1939 for the formation of the Women's Royal Naval Service.

Mrs Vera Laughton Mathews was invited to become Director WRNS on 31 March 1939 and accepted the appointment on 11 April.

The King approved the following Submission which was presented to him on 3 April 1939:

Lord Stanhope with his humble duty begs to inform Your Majesty that the Board of Admiralty have had under consideration the possibility of substituting women for men on certain work on shore directly connected with the Royal Navy; and it is recommended that, as in the last war, a separate women's service should be instituted for this purpose. It is submitted for Your Majesty's approval that this service should be called the Women's Royal Naval Service and that its members should wear a distinctive uniform, the details of which will be submitted to Your Majesty in due course.

The service would be confined to women employed on definite duties directly connected with the Royal Navy.

Should Your Majesty be pleased to approve these proposals, the Board of Admiralty would propose to appoint Mrs Laughton Mathews to be Director of the Women's Royal Naval Service. The Director will be responsible, under the Second Sea Lord, for administration and organisation of the Service, including the control of the members when off duty and the care of their general welfare.

It is humbly submitted that Your Majesty may be pleased to express your approval of these proposals.

Press Announcement

The King has given permission for the formation of

The WRNS unit on parade at HMS *Caroline, Belfast in 1941. (Lent by Mrs Baldwin)*

a Corps to be known as the Women's Royal Naval Service (W.R.N.S.) who will replace Naval Officers and Ratings in war-time on certain duties in Naval Shore Establishments. Training in peacetime will be given for some of these duties. Mrs Laughton Mathews, who served as an Officer of the W.R.N.S. in the late War and has since been one of the pioneers of the Sea Ranger branch of the Girl Guide movement, has been appointed Director of the W.R.N.S. and will enter upon her duties immediately. As soon as possible a booklet will be published detailing the conditions of service in the W.R.N.S. and telling candidates how to join. A copy of the booklet will be sent direct to all those ladies who have already sent in their names as volunteers for any Women's Service that might be formed under the Admiralty.

Advertisement For Port Superintendents – 14 April 1939

Applications are invited from women living at or near Portsmouth, Plymouth, Chatham or Rosyth for appointment as Port Superintendent of the Women's Royal Naval Service.

The Port Superintendent will be the officer responsible under the Director, Women's Royal

One of the many WRNS recruiting posters.

Naval Service, for the recruitment, efficiency, discipline and well-being of the women serving within their areas.

The WRNS officer's uniform made its first appearance at the National Service Parade, Hyde Park, on 2 July 1939, when Mrs Vera Laughton Mathews led eight officers.

Admiralty Regulations for the conduct of members of the WRNS were issued 23 August 1939.

Some of the earliest Wrens were recruited in the port areas and Mrs K. J. Wallace recalls the sinking of HMS *Courageous* on 17 September 1939:

> I volunteered for Service in the Portsmouth Command in 1938. . . I was called up and commenced duty on 14 September 1939 at HMS *Daedalus*. We had no uniform and wore mufti. On 19 and 20 September we were engaged in the sad duty of notifying next of kin of the casualties in connection with the *Courageous*, and received a written message from Rear Admiral Bell Davies in appreciation of our devotion to duty throughout the night.

Immediate proposals for the employment of WRNS personnel envisaged 126 officers and 1,475 ratings. An Admiralty letter was sent to all naval authorities at home on 20 September 1939:

> I am to acquaint you that requests for W.R.N.S. personnel to fill vacancies and release Naval personnel wherever practicable, to the maximum extent, for employment as Cypherers, Coders, Writers, Supply Ratings, Telephone Operators, Signallers, Motor Drivers, Messengers and Orderlies on the staffs of Senior Naval Officers of Shore Bases and Naval Officers-in-Charge of Commercial Ports at Home, should be forwarded to Admiralty through the Commander-in-Chief concerned.

Dr Genevieve Rewcastle, who was to be the WRNS Medical Superintendent throughout the War, was appointed in September 1939.

By agreement with the Admiral President of the Royal Naval College, Greenwich the first WRNS Officers' Training Course, which lasted two weeks, was opened on 30 October 1939.

In November 1939 the first WRNS personnel, Cooks, Stewards, Writers and Messengers, were sent to the Naval Base at Kirkwall in the Orkney Islands. In the same month a WRNS officer was appointed to

The WRNS officer's uniform has changed little in the intervening years – it was a good basic design.

the staff of Flag Officer-in-Charge, Northern Ireland and a unit of immobile Wrens employed in clerical and communications duties was established in Belfast.

Before the end of 1939 WRNS ratings were enrolled for intelligence duties as Linguists and Special Writers (Special Duties RFP); the training was carried out at the Campden Hill Depot.

In December 1939 there were 275 officers and 3,086 ratings.

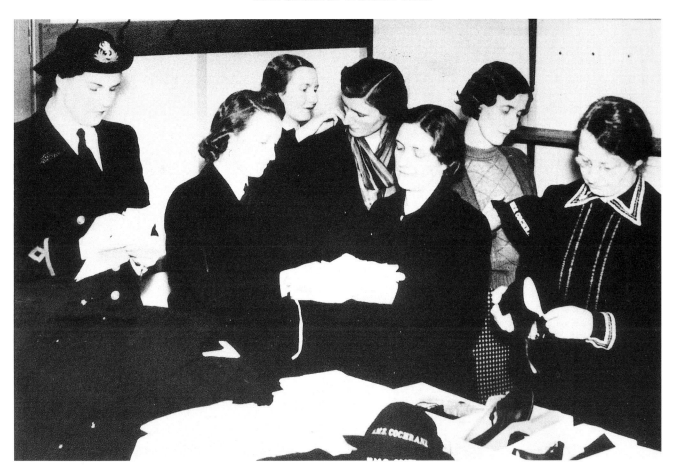

Above: *Uniform was not freely available in 1939 and Wrens were issued with a WRNS armband. Here, new entries are being kitted out at* HMS *Cochrane, Rosyth, in 1940.*

Below: *The Wren Despatch Riders were held in high esteem for their endurance and tenacity in overcoming any difficulty or hazard they met in the course of their duties.*

1940

The first organized general service training for WRNS ratings began in January 1940 when part of King's College of Household and Social Science, Campden Hill, London was taken over for the purpose.

The college catered for about 90 trainees in addition to the staff and London-based Wrens. Trainees performed the domestic work of the depot as part of their training and, whenever possible, Wrens destined for clerical or supply duties were given some professional training. In addition the first course for Special W/T Operators and R/T Operators started at Campden Hill in January 1940.

Approval was given for the formation of the WRNS Benevolent Trust and Wrens commenced working with the Naval Officer-in-Charge, Larne, Northern Ireland in January 1940.

Wren Despatch Riders, who were already employed at the Admiralty, took over completely from the Naval Despatch Riders in Spring 1940. Motor Drivers and Despatch Riders were, at that time, only recruited from women with previous experience.

The first Wren Degaussing Recorders were employed at Helensburgh in Spring 1940 where they controlled all the processes of range instrumentation for the ships on the range.

In March Her Royal Highness The Duchess of Kent became Commandant of the WRNS.

In May 1940 a Superintendent, Personnel, was appointed to be responsible for manning and recruiting. At the same time, a Chief Officer was appointed to deal with discharges, officer candidates, promotion boards, and was Staff Officer to Director WRNS.

Wren Cooks on the south coast were highly commended in June for their magnificent efforts in catering for the men returning from Dunkirk.

In June training was started for Stores (V); Special Operator Coder training and the Campden Hill Depot moved to Greenwich where some 230 new entry trainees could be accommodated.

During the invasion of the Low Countries the London Despatch Riders worked in eight-hour shifts day and night, carrying messages between the Admiralty and Embassies. The Netherlands Navy expressed appreciation of the work of the Wrens in a letter of commendation and thanks. They also won special praise for their work during the Battle of Britain when

Her Royal Highness, The Duchess of Kent, taking the salute at HMTE *Dauntless in March 1949. (Lent by Miss E. H. Scott)*

journeys often took two or three times as long owing to bomb damage.

After the fall of France three French warships came to Britain. The *Marshal Soult* and *Marshal Ney* became Royal Navy ships; the battleship *Paris* remained a French ship, although she was used as a depot ship for the rest of the War. Amongst the Wrens who worked aboard *Paris* was Mrs Dorothy E. Bennett who says:

I served as a Secretarial Officer from June 1943 until August 1945 on board *Paris*. Each morning the French flag was hoisted at the stern in parallel with the White Ensign. FS *Paris* became a depot ship for trawlers in the Plymouth Command. . . The ship's engines no longer functioned and large areas within her had been fitted out as workshops.

Wrens were employed as Writers, in the Signal department, as Officers' Stewards and as Boats' Crew operating small craft between *Paris* and the

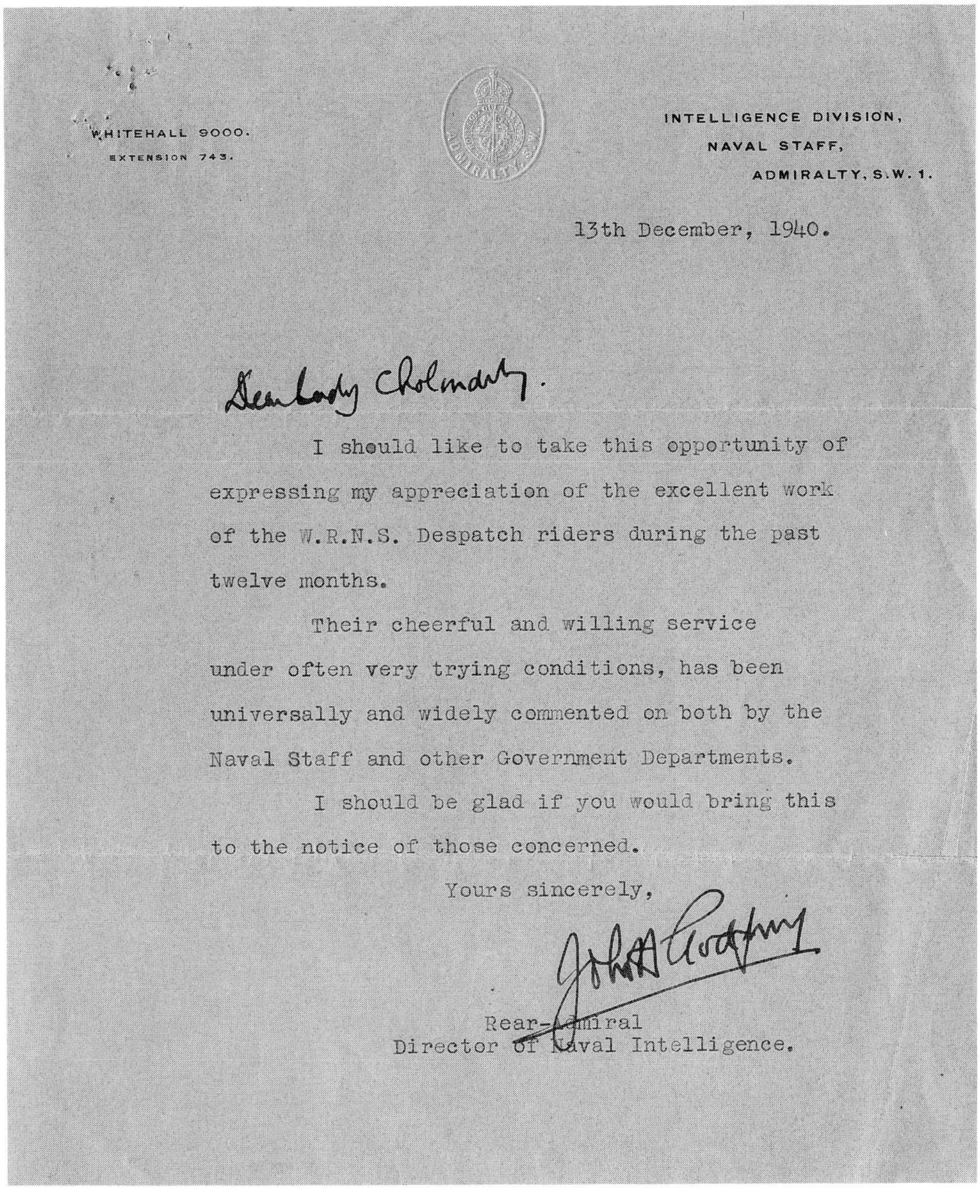

One of the numerous plaudits for Despatch Riders.

WHITEHALL 9000.
EXTENSION 743.

INTELLIGENCE DIVISION,
NAVAL STAFF,
ADMIRALTY, S.W. 1.

13th December, 1940.

Dear Lady Cholmondeley,

I should like to take this opportunity of expressing my appreciation of the excellent work of the W.R.N.S. Despatch riders during the past twelve months.

Their cheerful and willing service under often very trying conditions, has been universally and widely commented on both by the Naval Staff and other Government Departments.

I should be glad if you would bring this to the notice of those concerned.

Yours sincerely,

[signature]

Rear-Admiral
Director of Naval Intelligence.

trawlers. I, myself, lived ashore at the WRNS Officers' Quarters, but had a cabin aboard the *Paris* where I slept every third night as Duty Night Secretary.

Although officers and ratings were, from the outset, employed in Royal Naval Air Stations, the first of the more technical Air categories appeared with the battery charging and parachute packing tasks in the summer of 1940. The isolated situation of most of these stations precluded the use of immobile personnel and accommodation was provided similar to that built for men.

In the Autumn of 1940 a WRNS Press Officer was appointed to deal with publicity and for liaison with the Press.

Motor Transport Drivers from the Admiralty car pool carried out extremely hazardous work during the autumn when, in the incessant night air attacks on London, parachute magnetic mines were dropped. Many of these mines, scattered over a wide area, failed to explode. Small teams of experts had to be driven from incident to incident. Drivers took the teams as close to the mine as possible and waited whilst they defused them. There were a number of narrow escapes and only the roof of the car saved the driver from the mass of falling debris when a mine exploded killing two of the disposal experts.

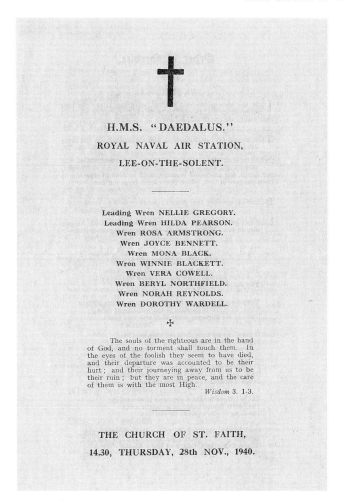

H.M.S. "DAEDALUS."

ROYAL NAVAL AIR STATION,

LEE-ON-THE-SOLENT.

Leading Wren NELLIE GREGORY.
Leading Wren HILDA PEARSON.
Wren ROSA ARMSTRONG.
Wren JOYCE BENNETT.
Wren MONA BLACK.
Wren WINNIE BLACKETT.
Wren VERA COWELL.
Wren BERYL NORTHFIELD.
Wren NORAH REYNOLDS.
Wren DOROTHY WARDELL.

The souls of the righteous are in the hand
of God, and no torment shall touch them. In
the eyes of the foolish they seem to have died,
and their departure was accounted to be their
hurt; and their journeying away from us to be
their ruin; but they are in peace, and the care
of them is with the most High.
Wisdom 3. 1-3.

THE CHURCH OF ST. FAITH,

14.30, THURSDAY, 28th NOV., 1940.

The Order of Service for the Wrens who died in the Mansfield Hostel. (Lent by Mrs Wallace)

In October immobile Wrens were recruited for employment in Londonderry, Northern Ireland.

Late in 1940 a two-week probationary period was introduced for ratings before they were kitted up and sent for further training.

On 14 September 1940 ten WRNS ratings were killed when their hostel in the Mansfield Hotel, Lee on Solent received a direct hit. Mrs K. J. Wallace has a vivid memory of this occasion:

In the summer of 1940, along with many others, our aerodrome was dive-bombed by the German Air Force and I can still recall the machine-gunning which preceded the bombs. . . One sad memory was when an unexploded shell from Portsmouth fell on the Wrens' hostel and exploded at the supper table killing ten of my colleagues and Wren Mona Black, a special friend, who came from a naval background.

Another Wren of great courage was a young 18 year old who worked in the Captain's office with me. She was a little Welsh girl from Cardiff and, during raids, she would sit amongst us reciting *Hiawatha* in her lilting Welsh voice, quiet and full of the music of the poetry – smiling around at us and exerting a calming influence never to be forgotten. Her name was Pat Humphries. . . tragically, she was a casualty when Mansfield Hotel was hit, and dreadfully injured. . . I mention her for her courage – a little unknown Wren.

Wren Plotters were employed at Dover in the summer of 1940 and by the end of the year 46 were employed across the four Commands. The category was to expand to include WRNS Plotting Officers.

In December 1940 there were 561 officers and 9,439 ratings.

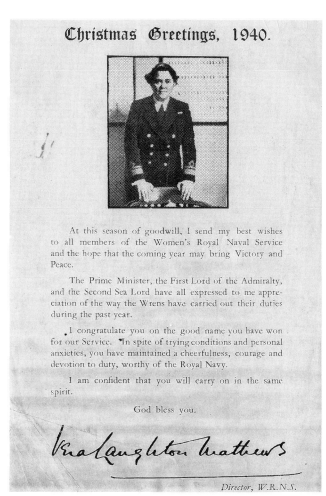

A christmas message from Vera Laughton Matthews to the Wrens.

1941

In January 1941 the first WRNS overseas draft sailed for Singapore. The draft consisted of Second Officer Betty Archdale WRNS in charge of 20 Chief Wren Special W/T Operators.

I was serving in HMS *Watchful* at Great Yarmouth when the Admiralty Fleet Order came out saying Wrens were wanted to volunteer to go overseas. As it happened Director WRNS was soon to visit Yarmouth. . . I asked her where 'overseas' was, to be told firmly by the Director, 'If you volunteer to go overseas Archdale, you volunteer to go overseas'. 'Yes, Ma'am', I replied. . . Not long after her visit a phone call came through. Would I go to Singapore? 'Yes, Ma'am', I replied – I was becoming quite well trained by this time. I also felt that

Singapore would be most interesting and pleasant, but what a pity we'd miss the rest of the war. How wrong I was.

Kitted up in a pretty terrible white uniform, including topee. . . off we went in a passenger liner. The unit was 20 Chief Wrens, who were all wireless operators, myself, and a first rate Nursing Sister, Lucy Travers.   **Betty Archdale**

Teleprinter Operator training commenced at Royal Naval College, Greenwich in January.

One of the first Battery Chargers to arrive at HMS *Kestrel*, Worthy Down in January was Mrs Mary Beck.

At last a chance to tell! So many books written about the WRNS in World War II and never a word about Battery Chargers or Boom Defencers. I joined in January 1941. . . a week at Princess Marina Barracks in Portsmouth and then to HMS *Kestrel* as the first Battery Chargers. We were joined by four more and that was our squad for two and a half years. Twelve hours on and twelve hours off – in a hut on the perimeter of the airfield. The 12-volt

The first draft to Singapore at WRNS Headquarters before leaving for the Far East. (Lent by Miss E. Archdale)

batteries had to be charged overnight ready for the trainee air gunners in the morning. . .

The countryside was so lovely, and how we laughed when Hitler claimed to have 'sunk' HMS *Kestrel*. Ralph Richardson and Laurence Olivier were there as pilots for a short time before leaving to make Ministry of Information films. Sir Laurence organised the concert party – six of us auditioned for him – and he obtained cancan costumes for us from Elstree Studios.

The Medical Officer at *Kestrel* finally decided that we must have other jobs – as the sulphuric acid fumes may have made us sterile! Happily we all had families later. At the end of 1943 I joined a new category – Boom Defence.

Expansion of Wrens employment gathered momentum during 1941 and 1942 and new categories gradually came into being.

Communications
Signal Exercise Corrector and Automatic Morse Transcriber were added to the branch, the latter category being trained by the GPO and the first qualifiers sent to the Royal Naval W/T Station, Flowerdown.

Bomb Range Markers at Skipness, Argyll. (Lent by Mrs Bain)

Parachute Packers
Although the proposal to employ women as Parachute Packers was taken in January 1940, it was not until early in 1941 that the first course was held at the Royal Naval Air Station, Yeovilton.

Bomb Range Markers
These ratings worked on ranges, entirely manned by WRNS personnel, attached to Fleet Air Arm Stations. Their duties were to assess, mark and plot aerial gunnery, practice bombing and rocket firing. They were trained to drive the crash tender and ambulance, to give directions to any aircraft off course or weather bound, and to deal with forced landings and picketing aircraft for the night. The peak number in this category was 354.

I was stationed at Skipness in Argyll for over 18 months from 1943 to 1945. Most of the time we had to make do with our own company – some 45 Wrens – with the occasional squadron party to break the monotony. Apart from two Cooks and two Motor Transport Drivers, we had no other categories serving with us and so had to do all the chores as well as our own duties. The event which stands out in my mind most clearly, was the pre-action training we gave to the squadron which was hoping to bomb *Tirpitz*. This was in March 1944 and we were particularly pleased when the Vice Admiral came up to Skipness to congratulate us.

Bomb Range Markers also served at Katukurunda, Ceylon and I served there from June 1945 until after VJ Day.   **Mrs V. Bain**

Meteorology

WRNS ratings were first employed on meteorological duties on Fleet Air Arm Stations during the summer of 1941. After six weeks' training they were employed in making routine weather observations, finding the height, quantity and type of cloud, recording temperature, wind force and direction, and estimating visibility.

Aircraft Checkers and Fabric Workers

These ratings identified and checked the movable fittings in aircraft according to an inventory and, later, they were employed on minor repair work on the aircraft's fabric.

Teleprinter and Switchboard Operators

An important event in the history of these categories occurred when, in 1941, the Admiralty took over the whole staffing of the Cambuslang Switching Centre for both telephone and teleprinter working. This was an instance of a naval establishment staffed entirely by WRNS ratings in the charge of a WRNS officer.

After six months as a Writer I requested to change my category to Communications. . . I was drafted to Westfield College for my course as a Teleprinter Operator and on completion went to HMS *Ferret*, Londonderry.

One day I was summoned by the Signal Officer and told that I would be working at the American Signal Station. . . young and rather quiet this came as a very great shock. . . I was happy with them and always treated with great consideration and respect. . . I have often wondered if, indeed, I was part of the Lend Lease Pact! I had to show my pay book to the Marine guard before entering and one new Marine, on looking at my pay book must have mistaken my initials, S. R. G. for some kind of rank for, as he handed my book back he said, 'You look mighty young to be a sergeant!'

Mrs S. R. G. Hague.

Vision Testers

This small category, never more than ten Wrens, was concerned with sensory motor apparatus, Later there were to be six Night Vision Tester Officers.

Personnel selection

By early 1941 the acute shortage of manpower gave rise to women being employed in the recruiting centres to carry out the whole range of recruit testing. WRNS ratings were given a special two-week course and rated Acting Petty Officer. They were also employed as assistants to the civilian psychologists in the Navy's new entry establishments. Some of them were, in time, to be promoted to officer rank and become Personnel Selection Officers.

Welfare and Amenities

Approval was given for the appointment of a WRNS officer in each of the five Home Commands as Assistant to the Command Amenities Liaison Officer. As the War developed a number of WRNS ratings were absorbed into these duties and were promoted to Chief Wren.

Mrs Daphne John, a Signal Typist, was seconded to the Command Amenities Liaison Officer in Portsmouth between 1941 and 1943. They had a store room for all the knitted comforts and, amongst other duties, arranged for men to go to Ditcham Park, a country house which was turned into a residential club where sailors could spend weekends.

Mrs J. M. Donaghy, on the other hand, started out as Writer to the Port Amenities Liaison Officer in Belfast and, in time became a Chief Wren Welfare Worker. Recreational and leisure activities were arranged for the escort groups which ranged from running a Navy club, to farmhouse weekends, football and cricket matches, the issue of games to the escort frigates, and free weekly trade cinema shows. The office was also a clearing house for SSAFA welfare visiting and reports. Applications from serving personnel and their families for compassionate leave, drafting near home for domestic reasons, release from the Service on compassionate grounds, financial hardship, housing aid etc., all needed investigation.

Experimental Gunnery

In January 1941 a WRNS officer was appointed to the staff of the Naval Experimental Officer at the Gunnery Experimental Establishment at Shoeburyness.

Her duties were the arrangement of trials, which involved detailing the gun, shell, charges, fuses, velocities and targets required, and the recording of the trial. She held the rank of First Officer and, as she was replacing a Gunnery Officer, she received specialist pay. During trials she had to be accompanied by a RA Officer as no woman was allowed to give the order to 'fire' or 'cease-fire'.

The first WRNS personnel arrived at Bletchley Park 15 February 1941.

I may be wrong, it's a long while ago, but I seem to remember that Station 'X' was the first name given to what later became PV. After the initial month's general training at the Royal Naval College, Greenwich, seven other ratings and I were the first to be drafted to Bletchley Park. Uniform was in short supply and we wore mufti with WRNS armbands.

On arrival we were allotted billets – also in short supply as the Foreign Office had taken the best. . . my Landlady's son slept in my bed when I was on night watch! When we were kitted out our tally band said HMS, and the only category badge was that of Writer which was very far from descriptive of our work.

We were all required to sign the The Official Secrets Act and I think that elsewhere it has been said that it was the best kept secret of the War. Winston Churchill sent a special signal to the Wrens of PV to say he was 'glad the Wrens were laying so well, without clucking'. I remember it raised our morale considerably.　　**Mrs P. Mack**

From March all recruits were entered as Probationary Wren and their category of work decided in the New Entry depot, instead of being nominated in advance of entry.

Custody and accounting of Confidential Books by WRNS officers was given official authority in March 1941 when an Admiralty Fleet Order stated, 'as Commissioned Officers they could be employed in these duties providing it was only in Shore Establishments'. They were not allowed to visit ships in the course of their duties.

On 21 April Admiralty administration of the WRNS was transferred from Civil Establishments 1 to the Naval Personnel Division of the Secretariat.

Matters concerning WRNS officers and ratings were referred to CW and N Branches. General matters were referred to the appropriate branches and departments concerned with naval personnel.

Westfield College, Hampstead opened as an extension for the New Entry Depot, Greenwich in April,

and professional training was commenced for Wrens Writer (Pay).

Mrs J. R. Dilks, who trained as a Writer (Pay) the following year, records her impressions thus:

I received calling-up papers to report to Westfield College for joint instruction with our male counterparts into the mysteries of 'mulcts of pay', 'hard-lying money', 'Grog' or 'Temperance'; punctuated by strident bugle calls. . . the clarion call of 'Gangway' whenever a dog appeared around a corner indoors – we knew a WRNS officer could not be far behind!

Cypher Officers' courses of two to three weeks' duration started at the Royal Naval College, Greenwich in May 1941, a second Teleprinter School opened in Dunfermline, and the first Wrens went to the British Admiralty Delegation, Washington, USA.

Cinema Operator

In July 1941 the Gaumont Picture Corporation started to train Cinema Operators. After a two week course the Wrens were drafted to shore establishments to show instructional films. Some were, in time, to serve overseas. Wren Cinema Operators were also employed with the Tipner Film Unit and the library, which filed all naval instructional films and was, at one time, entirely staffed by WRNS ratings.

In July a Chief Officer was appointed to the staff of Admiral Commanding Shetlands and Orkneys.

Mrs N. M. Hunt, who was a Steward, provides the following lively memory of this remote Command.

I commenced duty on 2 December 1942, reporting to Mill Hill, London where I was trained as a Steward. Eventually I was drafted to HMS *Cochrane II*. We all piled into the back of an Army lorry and set out for King's Cross Station. It was night time and, as we stood waiting to set off, a PO Wren came out and shone a torch on our legs to make sure we were all wearing regulation black lisle stockings and not the fully-fashioned kind.

After a short stay at Rosyth off we went to Orkney. I believe the ferry was *St Ninian*. She had a small gun up top, and a warmly clad naval person watched us all assembling along the deck. He had a fatherly expression – he knew it all and we were as green as grass.

On arriving at Lyness a PO Wren was there to welcome us with, 'You can leave your bags, you're

Wrens find a Hockey Match fun to watch even in the rain.

ss Aguila in which the first draft to Gibraltar was lost through a torpedo attack.

going to march up'. And so we did, with many naval faces gloating at these strange civilians who looked like fur-trappers. We marched up Haybrake Road to our new Nissen hut. Glory be, it was wonderful after our last posting. The hut was spotless, painted cream inside; we all had beds with new Navy anchor folkweave bedspreads, a small chest of drawers and, wonder of wonders, a warm radiator behind the bed.

Training for Wrens Stores (C), (V) and Naval Stores started at Westfield College in July.

Some Boat's Crew Wrens were trained as pilots to take ships across the Channel after D-Day.

Wireless Telegraphist training W/T 'W', started at Mathers, Dundee in August 1941.

I suggested to the Commander in Charge at Granton that we recruit some girls to train as operators. He finally gave way and eight local girls were recruited and trained, thereby releasing young lads for general service. News of our experiment reached Admiralty and the Director WRNS came up to Granton to see for herself. The Director must have been impressed for, within a short time, girls were being recruited and trained at Dundee.   Mr J. A. Dodds

When the age of acceptance into the WRNS was lowered to $17\frac{1}{2}$ I immediately applied to join. My father, who was at that time Telegraphist Lieuten-

ant in charge of Granton Radio Station, had told me of the acute shortage of operators, so I opted for the Wireless Telegraphy Branch.

We were billeted at Mathers Hotel, Dundee, but we received our training at the local Merchant Navy School. With three other girls I was sent to HMS *Osborne* at Culver Cliff on the Isle of Wight. We were to live in ex-coastguard cottages but worked in an old fortress, Fort Bembridge, a mile or so away. The fort was mainly inhabited by the Royal Artillery, with a small group of Royal Navy Asdic operators and Royal Air Force radar personnel.   **Mrs E. M. Foxon (née Dodds)**

The first draft of 12 Cypher Officers, 10 Chief Wren Special Operators and a Naval Nursing Sister left Liverpool for Gibraltar aboard the ss *Aguila* on 12 August 1941, as part of Convoy OG 71.

On 19 August the ss *Aguila* received a direct torpedo hit during an attack on the convoy by an enemy pack of submarines and Focke Wulfes. All the WRNS personnel were killed; a few survivors were picked up, including the Master of the ship.

Qualified Ordnance, Light Craft (QOLC)/ Qualified Ordnance, Combined Operations (QOCO)

In September 1941 the considerable expansion of the Coastal Forces, and the serious manpower shortage, led to Wrens being trained in maintenance of gun armament. The initial category of QOLC proved so successful, at Coastal Forces bases, that later a category of Qualified Ordnance was formed for duty at Combined Operations bases.

Mrs J. E. Pritchard was in Qualified Ordnance:

My category was rather the 'Cinderella' of the Service. I was in Qualified Ordnance, Combined Operations, known as QOCO. I was working on and behind gun ranges or, in the case of Royal Naval College, Dartmouth, in the armoury. There were only 200 of us in the Service and we all did our training at the Royal Naval Gunnery School on Whale Island. As far as I remember the course was about six weeks, during which we learned about all the guns, from pistol to pom-pom; how to strip them down, clean, oil and ensure they fired.

The maintenance of gun armament was only one of the many types of maintenance being taken over

during 1941 by WRNS ratings. They were employed on stripping, cleaning, painting, welding, fitting, turning, driving tractors and cranes, as electricians, sailmakers and carpenter's mates. From these beginnings the formation of many highly technical categories and branches for WRNS ratings and officers followed.

In September the second draft of WRNS personnel left for Gibraltar following the loss of the Wrens in the ss *Aguila* in August.

On 24 September 1941 WRNS officers' stripes became equated to naval officers ranks.

Boat's Crew

In October, in Plymouth, WRNS ratings were first trained for Boat's Crew duties. They were given a two weeks' course in general seamanship and, after training, either formed entire WRNS crews or took their place in mixed crews. Types of boats manned by WRNS ratings included 75-foot diesel harbour launches, various types of duty, mail, and liberty boats, and those used for transporting stores.

At Plymouth a WRNS crew under a Naval Petty Officer Coxswain manned an armed yacht, classed as a sea-going boat, and assisted in survey work. Survey boats required particularly skilful steering and manipulation of the echo-sounding gear. The Plymouth crew won high praise from the Admiralty Hydrographer who wrote: 'the results of their labours have been of high standard fully comparable with the work which would have been expected from a regular naval crew.'

Appreciation was also expressed by naval personnel returning from a long period at sea when met, sometimes in the worst of weather, by a WRNS Boat's Crew bringing out mail, or taking liberty men or cot-cases ashore.

Watches were usually 24 hours on and 24 hours off. In some of the larger boats the crew cooked, ate and slept on board. Peak numbers of the WRNS Boat's Crew category was 573. Memories by three ex-Boat's Crew Wrens give the flavour of the life they led.

I was a Boat's Crew Leading Wren. My initial training was given by officers from the Port of London Authority and Sir Alan Herbert on the *Water Gipsy*, whilst I was in the Minewatching Service on the Thames.

My duties in Newhaven Harbour were to service minesweepers and merchant ships with kite balloons, which were maintained and inflated by the RAF; to carry stores, mail and the Duty Officer out in the Channel to minesweepers which could not enter the harbour due to the mass of boats and ships which were tied up and preparing for the invasion; and to ferry personnel across the harbour.

I was also proud to be responsible for commencing the operation of one of our diversions. Just prior to the invasion I took out a buoy (containing a mechanism to appear as a fleet of ships on the German radar screens) into the Channel. It was then taken to the French coast by an MTB and left. Others were taken from ports on the south coast to the same area. This, reputedly, caused the Germans to divert troops to the wrong area, thus helping the invasion. **Miss F. Flowers**

In October 1941 I was interviewed by a WRNS Officer, having thought that I might be of use as a transport driver. I did mention, in a small voice, that I had done some 'messing about in boats', but had noticed that there was no such category. When called to report to Plymouth, it was as Boat's Crew.

Instruction was given in classes of four (I was one

of the first twelve) by a slightly aged Yeoman Signaller who viewed us with a somewhat jaundiced eye, apparently expecting little or nothing from us. We were drafted to Dartmouth to join a survey boat, where we were not made very welcome! The Captain's first orders to me, on a bitterly cold January day, were to scrub the boat from stem to stern, clean the bilges, clean and grease the chains, and point and splice every rope. Two ratings, awaiting draft, were real treasures, but the old stoker was a different proposition. Whenever I rang down to the engine room he would poke his head out from beneath the canopy and make his own decision. **Miss F. Hayes**

Miss Hayes was my Cox'n and I was the Stoker First Class aboard our 28-foot boat... We charted the map, fixed the buoy and the minelayer set the mines. These were laid in loops of five off Eddystone Lighthouse. We ate ship's biscuits, corned beef, and drank purser's coffee which we brewed on a primus stove in the forepeak. It is a miracle we four Wrens have survived all these years; we were always wet, poor 'Hayes' stood in the open at the helm in all weathers. I was lucky in my engine room apart, of course, when we moored, then I did my lighthouse climbing. We were in the Plymouth Blitz, bombed in Dartmouth, and machine gunned while moored

Boat's Crew at Portsmouth preparing to embark distinguished visitors for a visit to HMS **Dolphin**.

in Dartmouth Boat Float – they missed! I had just filled up with 50 gallons of petrol too! But we were at war, had a job to do and got on with it.

Mrs J. Shead

Torpedo Attack Assessors

In October 1941 two WRNS Officers were employed at Royal Naval Station Crail, Fife as Torpedo Attack Assessors. Their duties were to assess the angle of the torpedo when it struck the water, and the angle of the plane when it released the torpedo; calculating the degree of divergence and recording the result. Some of them later became Torpedo Attack Teachers. At the end of 1942 one WRNS Torpedo Assessing Officer was serving in South Africa, including duties afloat.

I joined the WRNS on my 18th birthday. After four weeks of preliminary training at Mill Hill, London, I was drafted to the Royal Naval Air Station Crail for work on the Torpedo Attack Teacher. With me were Wrens Barbara Tranter and Joan Selkirk.

In December 1942 I was drafted to the Royal Naval Air Station Hatston, Orkney, again on the Torpedo Attack Teacher. After operating for a while the Teacher was temporarily closed down. I understood that it was using so much electrical power that it was making it very difficult for the rest of the air station to operate.

Mrs R. D. Matthews

In October eight Cypher Officers deployed to Gibraltar aboard HMS *Malaya*.

Photographic Assistants

November 1941 saw a proposal which led to formal training at the Royal Naval Schools of Photography and Air Photography. In addition to standard photographic work the Wrens were often employed on developing and printing films for the Torpedo and Cine-Gun Assessors. Wrens were employed at the Aircraft Recognition Centre at Yeovilton, where they photographed models of enemy aircraft and prepared handbooks for instructional purposes.

In November the Officers' Training Course was increased to three weeks, and the intake of cadets varied from 20 to 30 every week. The aim of the course was to give a general picture of WRNS administration and of naval procedure and history.

Submarine Attack Teacher Ratings

At the end of 1941 Wrens were taking over the manning of the control table in the Submarine Attack Teacher in HMS *Dolphin*. As their skills developed they took over the submarine attack instruments, instructing submarine officers in the action, and eventually undertaking the attacks themselves.

Tailoress

Two Tailoresses were in service at the end of 1941 and the peak number was 25. They were employed in WRNS and naval new entry depots, having brought their skills from civilian life.

I volunteered as a Tailoress in 1941, but was put in as a Wren Steward at Mill Hill. I had worked for a time as a copy clerk in the Birmingham Tax Office, having been called away from my dressmaking business of 10 years. I was posted to Macrihanish in the wilds of Scotland. . . some time later I was sent to HMS *Cabbala* as a Tailoress – most enjoyable – and then back to Scotland to the naval base hospital Kilmacolm Hydro where I was in charge of all the linen.

Mrs L. Dougray

At the end of 1941 the Junior Staff Course was opened to WRNS officers. The following year Commander-in-Chief Western Approaches was to request the appointment of three WRNS Officers as Duty Staff Officers at Holyhead. This experiment proved successful and opened the way for further Staff appointments for women.

The National Service Act became applicable to women in December 1941. The three Women's Services were then regarded as part of the Armed Forces of the Crown.

In December 1941 there were 1,197 Officers and 22,898 Ratings.

1942

In January 1942 WRNS Officers were appointed for cash duties as Sub-Accountant Officers. There were to be 80 in all. Nine Parachute Officers, trained at Eastleigh, undertook secondary work in aircraft recognition and Intelligence – the following year parachute courses were superseded by the opening of the Safety Equipment School; and Special Duties Linguist training started at Wimbledon.

Second Officer Archdale and 30 WRNS ratings

were evacuated from Singapore to Colombo on 3 February 1942. When they had first arrived in Singapore they had found that their tropical uniform, which was made from the same material as the men's, was not very comfortable. With the connivance of the Admiral's wife they kitted themselves out in white frocks and no topees. They also wore slacks on watch which gave them some protection from insects. These unauthorised changes did not pass without comment from home as Miss Archdale recalls.

We were not in Colombo long and it was a sad time with little news of our friends. There was a letter from Director WRNS waiting when we arrived and I thought how nice of her to write and say she was

glad we'd got out of Singapore safely. How wrong I was. It was a real blast. What did Second Officer Archdale think she was doing altering the uniform? Only the Queen could do that.

By the summer of 1941 several WRNS Plotters had been promoted as Plotting Officers and, by early 1942, a considerable number were serving at home and overseas.

———

In March several changes took place.

The Central Depot, Greenwich closed and New College, Hampstead took over.

The National Institute of Medical Research, Mill Hill was requisitioned as the WRNS Central Training Depot.

The Royal Naval Accountant School, Highgate started professional training for the WRNS.

Wrens commenced maintenance on ships.

Wrens arrived in the Middle East and three WRNS Officers went to Canada to assist in setting up the Women's Royal Canadian Naval Service.

Some of the Chief Wrens who were at Royal Naval W/T Station Kranji, Singapore and were evacuated to Ceylon about five weeks before the fall of Singapore. They are wearing the unofficial, modified uniforms. (Lent by Mrs Gilbert)

Above: *In February 1941 Wrens were issued with bell bottoms.*

Right: *A Wren Ship Mechanic painting the new topmast on a landing craft.*

Wrens were involved in airborne radar research at RAF Defford.

In April the WRNS Benevolent Trust was inaugurated. During the war years the Trust was run by the Service for serving personnel in need. Its role was to change in later years.

At the end of 1941 a Senior WRNS Officer left to prepare for the employment of WRNS personnel in Alexandria. The beginning of 1942 saw the first of many drafts to Alexandria from where, in the summer, they were evacuated to Ismailia. From Ismailia some went to naval establishments at Port Said, Cairo and Suez. The unit in Singapore having been evacuated to Ceylon, went on to Kilindini where it was soon augmented by drafts from the United Kingdom. A small party was sent up to Basra, and the first draft for South Africa left in the summer.

WRNS personnel were serving overseas as follows:

	Officers	Ratings
Mediterranean, HMS *Nile*	49	286
Gibraltar, HMS *Cormorant*	39	43
Washington, HMS *Saker*	41	30
Kilindini, HMS *Tanga*	61	92
Nairobi, HMS *Korongo*	1	7
Cape Town, HMS *Gnu*	6	45
Simonstown, HMS *Afrikander*		
RATE Durban, HMS *Assegai*	6	200
Durban, HMS *Kongoni*	6	16
Wingfield, HMS *Malagas*	1	14
Basra, HMS *Euphrates*	1	8

The first six officers were appointed to the RNVR(S) Officers' Signal Course in May. Four of these officers were subsequently sent to Liverpool and, at the end of the year, Commander-in-Chief Western Approaches sent the following report:

Wrens J. Alldritt, J. Brown, and D. E. Henderson ashore in Durban. (Lent by Mrs Baldwin)

It is concluded that the experiment of training WRNS Signal Officers to relieve men has been successful. WRNS Signal Officers are capable of carrying out many of a Naval Signal Officer's duties, although they are handicapped by lack of experience.

The Petty Officer Wrens Mess at HMS *Assegai in 1944. (Lent by Mrs Baldwin)*

Eventually as many as 61 WRNS Officers successfully completed the course and several of them attained the rank of First Officer. Some officers acted as Port Signal Officers and Port Communications Officers.

Mrs M. Mears was a Signal Office Watchkeeper at Royal Naval Air Station Hatston, Orkney.

One of the most hectic periods I remember was the training of No. 8 TBR Wing in preparation for the bombing of the *Tirpitz*. As far as the SDO was concerned, that meant phenomenal activity because of the enormous number of deliberately fictitious signals being passed, on top of the normal heavy workload.

Topographical Duties

In May 1942 the Admiralty broadcast an appeal from the Inter-Service Topographical Department at Oxford for the loan of collections of photographs and odd snapshots of enemy and occupied countries. In June the Admiralty provided Wren Writers to supplement the civilian staff and later formed a Topographical category. Duties included indexing and filing, selection, captioning and copying.

Boom Defence

Wrens joined the Boom Defence organization in May 1942; numbers increased rapidly and, at peak, there were 194. At Felixstowe a Second Officer WRNS became the Control Officer. Eventually there were 14 WRNS Boom Defence Officers.

Boom Defence Wrens – who had ever heard of them? Tally band, HMS only. . . no category badges. . . nothing! In fact they appeared to be rather a rough, 'devil may care' lot, always seeming to be jumping in and out of Royal Marine lorries! Trips to Deal Station, unloading seemingly hundreds of boxes from the wagon trains. . .

Somewhere along the cliffs of Dover, at the end of a rough track, a large house had been requisitioned near to which two nissen huts had been built; these became our sleeping quarters. Closer to the cliff top stood six permanently erected brown tents, rectangular in shape, with one side completely open. Inside, clad in protective clothing, we girls would inflate large latex balloons from hydrogen gas cylinders and then, when each was fully inflated, came the dangerous part of the operation. To prevent the balloon from escaping, a heavy weight was attached and, to release it, one had to crawl underneath, praying with baited breath that friction with the canvas would be avoided – this could have resulted in immediate combustion and a resulting ball of flame.

The now inflated balloon required an attachment in the shape of a bottle of liquid nitrogen which, on impact with the ground, or overhead cables, would cause the whole thing to burst into flames. The robustness of the Royal Marines would then come into action. . . We struggled to hold back the rising balloon whilst the attachments were made and then, in a co-ordinated action, the balloon would be released; mercifully no-one was ever caught up and whisked aloft.

Mrs M. E. Cowans

At the end of 1943 I joined a new category – Boom Defence – a pseudonym. We assembled devices to disrupt, hopefully, the enemy output. Felixstowe was our base, on the estuary beach. We inflated latex balloons wearing flash hoods etc., to which was attached a sonde which had a timing fuse to enable it to reach a target in enemy territory. The labels were in various languages. The weather forecast was important – some days we could not operate so odd jobs, painting, assembling balloons, and making boxes in which to grow tomatoes were the order of the day!

Mrs M. Beck

Our job was, looking back, amazing! We were stationed on the barbed wire front at the Suffolk Convalescent Home in Felixstowe, and wore blue shirts or square rig and bell bottoms – also a white lanyard – as far as I know the only other category to glory in that privilege besides Boat's Crew! Our job was, when the wind and weather was right, to launch to Germany all sorts of anti-personnel bits and pieces – all on the end of large latex rubber balloons. These frequently caught fire landing us in hospital covered in gentian violet, or acriflavine, with no eyebrows and front hair – but nothing too serious.

Mrs A. Porter

In June 1942 waiting lists were so long that applicants for the WRNS were restricted to specific qualifications.

WRNS personnel were evacuated from Alexandria to Suez 30 June 1942.

I was in the fourth draft to Alexandria in April 1942 aboard the Troopship *Empress of Japan* (later renamed *Empress of Scotland*). I was an Ordinary Wren SDO and the ship was packed with about 5,000 Service personnel – the majority RAF and a few Army. We went round the Cape to avoid submarines, were disembarked in Durban, and continued our journey on board ss *Mauretania*. After 25 days at Alexandria we were evacuated with all the other WRNS personnel – Rommel's Army was advancing towards Alexandria. We were taken in cattle trucks to Ismailia – a long, uncomfortable journey and, later, by lighter down the Suez Canal to Port Tewfik. There we boarded the *Princess Kathleen*. Some Wrens remained on board for some weeks but I was fortunate to be one of ten watchkeepers who, after a few days, were taken by open lorry to Port Said. We were accommodated in part of a Roman Catholic convent, Le Bon Pasteur, on the edge of Arab town. I worked with another SDO Wren in the Signal

SDO Wrens at the Port War Signal Station, Port Said during the evacuation from Alexandria, 1942. (Lent by Miss C. M. Baker)

The Convent, Le Bon Pasteur, where Wrens were accommodated at Port Said. (Lent by Miss C. M. Baker)

Office in the dockyard of HMS *Stag*. We were transferred, about three months later, from the convent to the YWCA – a beautiful modern building on the sea front – I understand this move was due to the fact that plague was rampant in Arab town.   Miss C. M. Baker

Classifiers

In June Linguists and Special Writers became known as Classifiers. After an eight-week course, their duties were connected with the observation of the ionosphere and analysis of W/T transmissions. At peak there were 127 Classifiers, including some officers.

———————

In July the first WRNS W/T Instructors were appointed to the W/T Training School at Mathers, Dundee.

ss *Tamoroa* left the UK in July with the first draft of WRNS personnel for South Africa.

Some of the Wrens were in transit and, after a stop-over in South Africa, sailed for other destinations. Mrs P. Inverarity was one of these:

On hearing that volunteers were required for overseas I jumped at the chance. One hundred Wrens set out in the ss *Tamoroa*, a New Zealand ship. There were Army and Air Force personnel on board and it took five weeks of U-boat dodging, during which time we slept fully clothed wearing life jackets and clutching tin hats and gasmasks. We ran concert parties and even at rehearsal wore life jackets. At last we reached Durban and were housed in a hotel under very strict rules. No Wren was permitted to leave the hotel on her own; she had to have four companions. In the evening boy friends had to sign a book stating their rank, regiment, ship etc. Word got around amongst the lonely servicemen that they could go to the hotel, sign a book and be given a dancing partner for the evening. The Officer in Charge soon let them know this was not true.

All of a sudden eight Wrens were detached from the main group. All Teleprinter Operators. We were told we were going to Basra. We sailed in a Dutch ship, with troops for Bombay and the Far East. Two weeks at the United Services Club in Bombay – in great luxury – early morning tea and fruit followed by a huge breakfast. At last we were told to pack and were taken to the docks where we went aboard a very dirty and extremely hot British India ship. Mice, beetles and other unwelcome creatures were already in residence in the cabin allocated to us – all eight in one cabin. Finally we arrived in Basra to be greeted by the Iraqi Army playing excerpts from 'Rose Marie' on the quayside.

AA Target Operators

This category, started in the summer of 1942, contained 22 Wrens at its peak and one WRNS officer.

The Wrennery at Cape Town. (Lent by Mrs Baldwin)

Chief Officer Ann Rogers and the WRNS Officers, HMS Gnu, *in 1944. (Lent by Mrs Baldwin)*

47

Naval Censorship and Fleet Mail
The Naval Censorship branch, in which WRNS officers were already serving, was extended to include the appointment of WRNS officers as Fleet Mail Officers.

I joined the first Mobile Censorship Party early in 1942. As a group, when not travelling round the country, we worked in the Censorship Headquarters in Holborn. Eventually there were three Mobile Parties. **Miss A. Winser**

I joined the WRNS in 1940 in Belfast. After a couple of years as a Coder I attended the OTC at Greenwich and became a Censor Officer, mobile, attached to Intelligence. After some time as a Censor Officer I was sent on a Fleet Mail Course at the GPO Mount Pleasant. I remember seeing a vast mountain of mail bags waiting to be sent to the besieged island of Malta. To complete the FMO training we were sent in pairs to work at a large Fleet Mail Office. I was delighted when I was sent, with Third Officer Celia Clabby, to Scapa Flow. We were based at Lyness and worked on board the Depot Ship HMS *Dunluce Castle*. It was cold and bleak at Scapa but an enjoyable change from

London, quite a sight to see the many large ships which came and went.

After Scapa I was appointed to Sheerness, HMS *Wildfire*, I remember that the motor launch in which we delivered and collected mail, for censorship and despatch, had been to Dunkirk and had bullet holes to prove it. In 1944, at my request, I returned to Belfast. Before the D-Day landings we had long queues of US Servicemen waiting to embark; we spoke to many of them, and took messages and last minute notes from them. It brought home to us very clearly what these men were facing, and it was a very tense time for all of us. **Mrs P. Lachlan**

ARP Officers
The first appointments of WRNS officers as ARP Officers were made in 1942. Eventually there were three of them and one was appointed as a travelling lecturer.

Orthoptists
In the summer of 1942 the Orthoptic Training Centre was opened at Royal Naval Air Station, Lee on Solent and two WRNS ratings, qualified in civilian life as

Mail Staff Wrens prepare deliveries for HM Ships.

Dame Vera Laughton Mathews launched HMS Wren *on 11 August 1942. The ship was later adopted as the WRNS ship. (Lent by Mrs Broster)*

Orthoptists, were promoted to officer rank and appointed for duty. This small officer branch eventually increased to six.

———

A Chief Officer WRNS was appointed to the Fleet Air Arm Drafting Office at Lee-on-Solent, in the summer of 1942, to cope with the expansion of specific WRNS Fleet Air Arm categories.

Cine Gun Assessors
A few WRNS ratings were employed in this capacity at Royal Naval Air Station, Yeovilton in 1942, and were soon in evidence on all the larger Air Stations.

They were taught to assess, from pictures taken by cine cameras, the accuracy of a fighter pilot's aim. They plotted their findings on a graph and explained the findings as necessary. At Yeovilton a WRNS officer was in charge of the Cine Gun Assessing Department, combining her duties with those of an instructor to naval and WRNS personnel. The category peaked at 87 WRNS ratings and four officers.

———

The Dieppe raid afforded a striking example of the confidence placed in WRNS Writers. The Naval Force Commander in charge of the operation asked for a WRNS Writer to act as his personal writer. This rating, with a small team of Wren Writers, dealt entirely with the written work connected with this planning, made up operation orders, and bundles of escape packages, and attended briefing meetings to take verbatim reports of proceedings.

Despatch Riders

A Special Order of the Day was issued in Portsmouth General Orders on 22 August 1942.

ORDER OF THE DAY   22nd August 1942
Ten W.R.N.S. Despatch Riders have, during the last fortnight, covered 10,000 miles and delivered several hundred immediate and important despatches, without a hitch. On the night of 19th, Ldg. Wren Tustin led convoys in thick mist and over strange roads to their destination, and P.O. Wren Harris did valuable service in carrying a Staff Officer over dark and difficult routes. Both these Wrens were 21 hours on duty without a pause. P.O. Wren Harris had previously covered 250 miles in $7\frac{1}{2}$ hours running time, 100 miles being in the dark

and, apart from 2 hours sleep, was, on that occasion, on duty for 26 hours.

Ldg. Wren Fergusson made a trip of 200 miles over strange roads in the dark in $8\frac{1}{2}$ hours with a despatch and, afterwards, completed the 300 mile trip to Dover and back in 9 hours running time.

Wren Steel completed a 200 mile journey to Plymouth in $5\frac{1}{4}$ hours, and Wren Marsden the same trip in $10\frac{1}{2}$ hours, at night, despite a puncture and having to use a torch for 20 miles after her lights failed. To both these Wrens the road was strange and included crossing Dartmoor.

Other similar difficult journeys, many of them at night, were accomplished swiftly and surely by other Wren Despatch Riders.

This is a record of achievement and duty well done that I feel should be known throughout the Command.   Admiral W. M. James

Wren Despatch Riders pass in review before Director WRNS. (Lent by Miss Neale)

Flat sailor-type caps replaced the waterproof twill hats in August 1942.

In August W/T 'W' and V/S training transferred to HMS *Cabbala*.

I joined the WRNS in 1942, having volunteered in 1941. I was sent to Mathers Hotel, Dundee to train as a Wireless Telegraphist. It was part of HMS *Cressy* and we trained at the Merchant Navy Signal School. Halfway through our course Mathers closed down as a WRNS Signal School and we transferred to Lowton St Mary's, near Warrington, to a new camp which had just been built – HMS *Cabbala*. We were the first Wrens there.

Mrs E. Nelson-Ward

The first two WRNS Meteorological Officers finished their training at Greenwich in August and were appointed to the Air Stations at St Merryn and Hatston.

Early in 1943 Rear Admiral Naval Air Service reported that, 'WRNS Officers have in general justified their appointment for meteorological duties, and the scheme is worthy of expansion.' The officers worked in watches with junior naval officers forecasting and warning Flying Control of any sudden or unexpected changes in the weather. They were trained to forecast flying conditions for the next twelve hours and to provide weather information for special routes. Eventually WRNS personnel were at most of the Fleet Air Arm Stations at home, and three officers and eight ratings served in South Africa.

Minewatchers

Ninety Wrens were on duty during air raids in blast proof observation posts. From these they watched given stretches of water, took bearings, and reported the position of any mine which fell.

On 2 September 1942 I reported to the Mill Hill Training Depot. From there several of us were drafted to Ormond Lodge, Richmond, Surrey to be trained as Minewatchers. Every evening we were deposited at different posts along the Thames as far as Westminster Bridge. In the event of an air raid we had to run to the look-out pill box and take bearings of any mines that dropped into the water so that they could be swept as soon as possible. When there were no air raids we slept in a variety of strange places; schools, hospitals and even Spalding's house at Putney where large numbers of tennis balls were stored. . . In the day-time we attended lectures, fire-watching, squad drill and so on. The highlight was going on the river with A. P. Herbert in the *Water Gipsy*, when we were instructed in tying knots and navigation etc.

Miss H. Elliston

NCS Routing Officers

In September 1942 a memorandum from Director of Trade Division stated that 'owing to heavy demands made from time to time for officers with routing experience for appointments abroad etc., it has become necessary to consider the possibility of diluting the Routing Staffs at UK Ports with WRNS Officers.' The duties comprised the correction of charts etc., and general clerical duties but, at first, 'excluded direct dealings with Masters in so far as the routing instructions were affected.'

In a few months there were at least 17 WRNS Routing Officers at home, and two at Algiers and Port Elizabeth respectively.

A WRNS Boarding Officer returning ashore after handing over sailing orders.

Drawing Duties

Wrens were already being employed in small numbers on diagram drawing, tracing etc., and in September they were formed into the Drawing Duty Category. There were 85 at peak, including some serving overseas in HMS *Bherunda*, Colombo.

Wrens were re-established in Alexandria, Port Said, Ismailia, and Cairo in September.

Torpedo Wrens

In the summer of 1942 the shortage of Seaman Torpedomen was serious and led to Wrens being trained to replace them. Initial training was started at the Government Training Centre, Hounslow, and duties included the care and maintenance of torpedoes and depth charges, the electrical fittings on Coastal Force craft, stripping, cleaning and replacing

A Torpedo Wrens Working Party.

delicate mechanism and all other aspects. A large number qualified as Leading Torpedomen and the first Wren (T) promoted to officer rank was trained as a Torpedo Attack Teacher. Torpedo Wrens worked with Special MX Section at HMS *Vernon* and, in October, five Wrens took over the Submarine Attack Teacher, Dundee.

I was a Leading Torpedo Wren (L). Having finished my course I was drafted to HMS *Vernon* and found myself in MX. The work there was 'hush-hush' until 10 December 1945 when details were published in the Daily Papers, and an article by Richard Dimbleby appeared in *The Listener*.

The work was connected with experimenting and making the electrical sections of magnetic and acoustic mines used in submarine warfare. Some of the Wrens went to HMS *Birnbeck* (the old pier at Weston-Super-Mare) where the mines were dropped by the RAF and recovered by the Royal Navy.   Mrs M. Powell

Radio Mechanics

Radio Mechanics had appeared early in 1941, but the category officially came into being with the commissioning of HMS *Ariel* as a training establishment for Radio Mechanics. Until this time some Wren Radio Mechanics had been employed at the ports servicing ships' W/T sets, but the majority were with the Fleet Air Arm servicing aircraft W/T and radar sets. Later this was to be extended to Coastal Forces craft equipment. Peak numbers reached 1,132.

Miss M. E. Callanan who joined in 1943 spent the first four months of her professional training at the S. E. Essex Technical College to be followed by a specialist course in radar mechanics at HMS *Ariel*, Warrington. Next came further courses at Portsmouth and Lee-on-Solent. After about seven months' training in all, she was sent to HMS *Drake* where she was attached to the Gunnery School. Her duties were associated with maintaining equipment for training purposes, and setting up project equipment for radar research.

Dome Operator

This specialized branch of the Cinema Operator category was formed to operate the domes used at most gunnery schools for training purposes. Numbers never exceeded 13.

Radio Mechanics in their workshop.

Maintenance (Air)

By November 1942 so many WRNS ratings were being employed on maintenance work that it was decided to form a branch to be known as Maintenance (Air). This then embraced the Battery Chargers, Aircraft Washers, Sleeve Target Packers, Plug Cleaners, Electrical and Air Frame ratings, and all unspecialized air maintenance work.

Gunnery Control

In addition to gun maintenance, Wrens were employed on gunnery sites as Analysers (S), Range-finders (S), PO Sights (GD), and Rangetakers at AA Targets (S). By the end of 1942 all the gunnery associated duties were combined into Gunnery Control. There was also a small category known as Gunnery Experimental Assistants.

Hairdresser

Hairdressers were introduced to overcome the lack of facilities in isolated WRNS units, and brought their skills with them from civilian life. At peak there were 148.

In December 1942 there were 2,270 Officers and 38,145 Ratings.

Mrs Eleanor Roosevelt inspected Wrens on parade in Liverpool on 8 November 1942. (Lent by Dame Nancy Robertson)

Their Majesties visited HMS Caroline, *Belfast in 1942. Chief Officer Rogers (far left) was the first Officer-in-Charge WRNS, Northern Ireland. (Lent by Mrs Baldwin)*

Her Majesty The Queen inspecting Royal Guard of WRNS ratings in Liverpool on 16 November 1942. (Lent by Dame Nancy Robertson)

A group of WRNS Officers at Westminster Abbey in 1942. Left to Right: 2/0 Jones, 3/0 Hughes, 3/0 Langley, 1/0 Woodhouse, 1/0 Carter, 1/0 Hardy, Sister Myers, 2/0 Kellard, 3/0 Stephens, 3/0 Geddes.

1943

The beginning of 1943 saw the first occasion on which WRNS officers accompanied the Prime Minister to a conference with other war leaders. In January, 12 Secretarial Officers went to Casablanca to be followed, in May, by 21 Cypher Officers attending the Washington Conference. During the voyage they worked in three watches. In August, 30 Cypher Officers were to attend the Quebec Conference and in November, 31 went to Cairo – some of whom went on to Tehran.

During 1943 there was a great expansion in the size of the WRNS unit in Orkney and Shetland sub-command. Doubts had been expressed about employ-ing women in inaccessible areas but, eventually, a unit of 65 officers and 615 ratings was established at Lyness. Over 230 worked at Wee Fea as Watchkeepers, where most of the communications was in the hands of Wrens. Large numbers of Cooks and Stewards worked at Lyness and the Pay and Supply branch was staffed largely by WRNS personnel.

After arriving at the jetty at Lyness, and being marched up the road by a PO Wren we arrived at our new Nissen hut. From the window, beyond the fuel tanks, was a perfect back-drop – a huge hill called Wee Fea. It changed colour all the time as the weather altered.

We were given 12 or more officers to care for, and had taken over from male Officers' Stewards who had been drafted to HM Ships. One of the Stewards took us for a walk to get our bearings and the first thing he did was to show us the cemetery – telling us that several men had gone mad with the isolation of the place. . .

Every morning my duty was to rise early, go

*A WRNS Boarding Officer leaving a merchant ship after
seeing the Master with confidential orders.*

across to the officers' cabins where we made and served morning tea, collect their shoes and suits, which we polished and brushed; return to our quarters for breakfast and then back to make beds and clean up. During off duty time we went ship visiting and attended dances. There were trips to Stromness where I bought a beautiful pair of green silk pyjamas for 18/11d and a hand made jumper for 7/6d. On Lyness one could buy Palmolive soap and in Stromness one could purchase sheer black fully fashioned stockings. . . Going on leave was a long, tedious journey, where the only good thing was the wonderful Salvation Army who ministered piping hot tea and meat pies to us.

Mrs N. M. Hunt

I have never seen a mention of HMS *Fox* and yet there were quite a number of Wrens stationed in Lerwick. My time in the Shetlands was the greatest experience and, had it not been for the War, the most enlightened time of my life. . . My husband and I go north every year and there are still many folks, now elderly, who call me 'The Peerie Wren'. I remember when word got around that we were leaving Shetland, one old lady from the Widows' Homes said how vexed they were to lose us as we had been very well behaved.

Mrs C. Fraser

Boarding Officers with NCS

Although when WRNS officers were first introduced to NCS they were not allowed to board ships, the increasing shortage of manpower led to a review and the Director of Trade Division proposed that 'especially active and athletic WRNS officers' might undertake duties as Boarding Officers. After eight weeks' training they went out in all weather to board merchant ships to give and, where necessary, explain confidential orders to Masters. The following tribute was received from the Commanding Officer at Milford Haven:

The fact that WRNS officers could service a weather-bound convoy, and sail the bigger ships in the gales that were blowing last week, shows conclusively that WRNS officers are fully capable of taking over this branch of NCS work.

In January the Ear, Nose and Throat Hospital, Golden Square, London became an Overseas Holding Unit.

In 1942 I volunteered for overseas service and on 3 March 1943 I was sent to London, one of a draft of 100 Wrens quartered in what had previously been a hospital in Golden Square. We were given tropical kit and lectures, and drilled outside in the square by Chief Wren Regulating Beaumont-Nesbitt, our capable CPO in charge of the draft. During the air raids on London we had to go to the other side of the square to an air raid shelter – so we wondered if we would ever make it overseas. However, one very dark cold morning, before dawn, in strict silence and secrecy we climbed into open lorries and were driven to St Pancras Station for an unknown destination. We eventually arrived at Liverpool and marched, through the bombed and rubble-strewn streets, to the docks where we embarked in a huge troop transport ship the HMT *Monarch of Bermuda* – destination Cape Town.

Mrs D. E. Baldwin

The Chief Officer WRNS on the Staff of Commander-in-Chief, Levant, was promoted to Superintendent on 22 February. Her area included all the North African and Persian Gulf units.

I was sent by sea to Bombay and then up the Persian Gulf to Basra. There were only some half dozen Wrens who, I think, had been meant for Singapore but were halted in South Africa and then sent up to the Gulf. We were on the Staff of the Senior Naval Officer in the Persian Gulf. The Wrens were Writers and, as it was such a small unit, I did secretarial work.

One day I was taken for a drive to the alleged Garden of Eden. I still have a slither of wood a small boy cut from the Tree of Knowledge for me.

Miss Betty Archdale

Early in 1943 the War Cabinet approved the employment of WRNS Cypher Officers and Coders on board 'Monsters' – troopships.

The ratings were signed on as part of the ship's company. Teams were made up from two to four Cypher Officers and three to six Coders. Officers usually worked 24 on and 24 off, whilst ratings worked in three watches. All were volunteers. Some did short trips whilst others were at sea for up to seven months. By the end of 1943 seven troop-carrying ships carried WRNS Cypher Officers.

The following memory was printed in the February 1960 issue of *The Wren*.

6 June 1943, was an exciting day for me. I was bound for Halifax, Nova Scotia as a Wren Coder on board the troopship *Pasteur*. Two Cypher Officers, one Second and one Third Officer, and two Wren colleagues comprised the complement. The 'Monster' trips, as they were called, had started in the spring of 1943, and the Liverpool Wrens had the *Pasteur* for their own special ship... The passengers were 2,000 German prisoners, mostly Rommel's Afrika Corps. They were housed in the bowels of the ship and we were to see very little of them... we decoded quite a lot of traffic. The *Queen Mary*, also on troopship duties bound for New York, was following us for part of the voyage and we took a great interest in her signals... One of the most important was the daily U-boat disposition. The crossing was made in four days... we arrived at Halifax in the early evening, anchoring in the harbour, and it was simply unbelievable to see lights blazing everywhere – no black-out... we spent ten days in Halifax before setting sail for Liverpool with a very full passenger list – 5,000 Canadian Air Force personnel... Arriving in Liverpool we had a few days in quarters and then we were off again. **Edith A. Brown**

On 2 March 1943 troopship *Empress of Canada* left Durban for the UK carrying six Chief Wren W/T who had been evacuated to Colombo from Singapore. At 2300 on 13 March the *Empress of Canada* was torpedoed. All the Chief Wrens were saved, having been rescued by HMS *Boreas*.

We had arrived in Kilindini Harbour, Mombasa on 3 May 1942. Our watchroom was constantly plagued by masses of flying insects (including praying mantis) and bats... we were issued with hideous thigh-length canvas leggings and cotton sleeves to be worn after 6 p.m. as an anti-malarial protection. Needless to say they didn't stay on long after our leaving quarters. By 1943 more and more Wrens were arriving from the UK (all different categories) and we originals were drafted home in three parties.

The first draft was torpedoed when the *Empress of Canada* was sunk on 14 March in the southern Atlantic by the Italian submarine *Leonardo da Vinci*. The six Chief Wrens were Bonner, Finch, Dart, Gadd, Gray and Kopsen. All were saved although some were afloat for up to four days on rafts and lifeboats. **Mrs J. Dinwoodie**

Freda Bonner, who qualified as a doctor after the war, was one of the Wrens on board the *Empress of Canada*. In the June 1960 issue of *The Wren* she wrote an article about the sinking of the troopship.

The evening of 13 March passed as usual except that the ship was vibrating noisily and obviously her engines were going flat out... we were all in our cabins by about 10.30 p.m.

Shortly after 11 p.m. we heard a dull thud, far, far, below us. We knew at once that we had been torpedoed and a few minutes later the ship stopped dead. No alarm sounded; there were no lights, even the emergency system failed – a brilliant, or freak, shot had hit us amidships in the engine room. At first there was an uncanny silence and then we heard the urgent footsteps on the stairs. We Wrens were all together in one cabin. In a few minutes we had put on our coats and lifejackets. I put my hand into my dressing case to pick up some money, but immediately thought, 'I shall never need this again', so left it and, instead, picked up a packet of cigarettes – I do not smoke... When we found our boat station, our boat was overturned in the water ... the ship had heeled over, the decks were at a sharp angle... half the lifeboats could not be launched or had overturned. Naval officers took charge throwing rafts and floats into the sea, letting down ropes, getting off refugees and the Italian POWs who jumped into the sea. The Italian Submarine Captain observed the Geneva Convention ruling to allow time for the passengers and crew to abandon ship then torpedoed her again... As we were all pushing away from the ship, suddenly, in front of us, the seas opened and a huge black monster came up from the deep, it was the submarine. At gun point the Captain ordered the nearest lifeboat to come alongside, from which he took the only Italian Officer...

It was about two hours later that the *Canada* suddenly upended and, with a dreadful, awful roaring noise, plunged down into the depth of the sea. This dreadful, awful, roar remained with me night after night for nearly a year... It was a long night... there were many cries for help... at long last the dawn came. All around us were rafts with red sails, boats with red sails, Carley floats and still many people in the sea, their lifebelts having supported them all through the night... there were also the fins of sharks and barracudas.

On the second afternoon a Sunderland flying boat flew over us, we knew then that rescue was on the way. On the fourth evening, as the sun was setting, we saw HMS *Boreas*. We watched her stopping over and over again to pick up survivors, and soon it was our turn. Every stop the Captain made endangered his ship. . . two days later we were landed at Freetown. . . it was a sad accounting, all six Wrens and the ten women refugees had been saved, but the cost had been high. Over two hundred had been lost; these losses would have been much higher if it had not been for the Royal Naval officers who gave their lives 'doing their duty'.

Finally the *Mauretania* brought us to England. From there I had to get to my home in Dublin. By accident the porter at Euston put me in the wrong train. After three changes an inspector told me that I must change again at Bangor. I was tired and exhausted and, in tones of despair, I said, 'Must I change yet again?' A woman beside me said, 'You're lucky to be travelling – don't you know there is a war on?'

HMT *Monarch of Bermuda* sailed from Liverpool for South Africa on 13 March with a large draft of WRNS personnel destined for Cape Town, Simonstown and Durban.

Having arrived at Liverpool and embarked in the *Monarch of Bermuda* we joined a convoy of about 30 ships and sailed into the unknown. We were instructed to sleep in bell-bottoms for the first few nights as it was at that time the height of the German U-boat campaign. Protected by cruisers and destroyers we zig-zagged our way across the Atlantic until, one day, the green and rust of Africa loomed out of the sea and we sailed into Freetown.

After Freetown we sailed almost to the Argentinian coast, still not knowing our destination, and then one day, nearly four weeks out of Liverpool, we sailed into Table Bay at Cape Town and beheld the majestic sight of Table Mountain.

Mrs D. E. Baldwin

Five WRNS personnel were killed and thirteen rescued when WRNS quarters in Great Yarmouth suffered a direct hit on 18 March.

At 0628 an enemy plane dropped six HE across the southern part of the borough. The worst incident was the demolition of large houses which were occupied by WRNS personnel. Thirteen trapped Wrens were rescued alive and great credit was due to the rescue parties for their excellent untiring work. Five Wrens died, including the Quarters Officer.

Mrs M Brown

Fleet Mail Officers

By March a special course was being held for WRNS Fleet Mail Officers, and by June there were 170.

I joined in November 1942 as a Fleet Mail Clerk. After initial training I was drafted to HMS *Monck* at Largs in Ayrshire. It was a Combined Operations establishment and some parts of 'Operation Torch' had been planned there. Lord Louis Mountbatten visited us, amidst great secrecy. After a short period I was drafted to HMS *Gannet*, Royal Naval Air Station Eglinton, Northern Ireland. Their Majesties and The Princess Elizabeth visited the station and I also witnessed the official surrender of the U-boats, in 1945, to Admiral Sir Max Horton at Lisahally, County Londonderry.

Mrs B. Millott

Air Mechanics (O) (A) (E) (L)

HMS *Fledgling* at Mill Meece was opened in March as a training centre for Air Mechanics.

The Air Mechanics (O) stripped, cleaned, tested and re-assembled aircraft armament; (L) tested electrical fittings, did minor repairs, overhauls and signed airworthy certificates; the (A) and (E) did the same service for airframes and engines. Peak numbers in 1944 totalled 1,581. Three Air Mechanics served overseas 2 (L) and 1 (E) – all at Roosevelt Field, USA.

I was in the first class of Air Mechanics (Engines) at Mill Meece. Classes were also started for Airframes and Electrical. We were taught basic use of tools, the various parts of aircraft engines – all in about five months. *Aeroplane* magazine published photos of the first Wrens to be actually trained as Air Mechanics.

Our classes were sent to two Fleet Air Arm Stations at the end of the course – Crail in Fife, and Yeovilton in Somerset.

Mrs L. M. Farrell

I was one of the first Wrens to start training at Mill Meece. Our first priority was to clean – and I do mean clean – the various huts, which had been left in a filthy state by the American Army. . . we went straight to Mill Meece with no initial training.

Wren Air Mechanics under training at Morning Divisions.

There were munition factories next door to the camp and the girls working there often used the same buses. I felt sorry for them – hair and skin coloured bright yellow from the work they did. On completion of the course I went to Northern Ireland – certainly the men in the hangars wondered what had hit them, but soon accepted us as part of the team. **Miss M. Somerset**

I joined in 1944 and went to Mill Meece for training as an Electrical Air Mechanic. The basic theory of electricity was taught to us by a 'Schoolie'; practical work was taught and carried out in the workshops; then the basic training was applied to actual aircraft and we learned how to inspect them, and then sign the Form 700 certifying that the aircraft was fit to be flown. The final examination was carried out on the airfield, finding faults with real aircraft. **Mrs F. J. Wakem**

Soon after the War Cabinet approval for communications personnel to serve afloat, approval was given in May for Writers to serve on board troop transports. They, too, were volunteers and they assisted the Officer in Charge, Naval Draft.

Mrs C. L. Pearn was a shorthand typist who was drafted to Larne, Northern Ireland in spring 1943.

I was sent to the Captain's office of the Naval Officer in Charge, HMS *Racer*, but had only been there for about three weeks when the Commander-in-Chief, Western Approaches, Admiral Sir Max Horton, decided to make Larne the headquarters of the Training Captain, Western Approaches. He arrived in his ship HMS *Philante*, required a good shorthand typist and I was the lucky one chosen.

Class El after 'Passing Out' at HMS *Fledgling, Mill Meece in August 1943. (Lent by Mrs Farrell)*

This was at the height of the Battle of the Atlantic and our function was to take escort groups, submarines and aircraft carriers on three-day exercises in the Irish Sea.

Net Defence

This category was formed in May 1943 from Maintenance Wrens, and was concerned with oiling and cleaning nets used in defence of quays and ships. Peak number was 72.

DEMS Inspecting Officers

Two WRNS officers were given a fortnight's training at the Gunnery School learning to strip and reassemble all guns from rifles up to six-inch, and familiarize themselves with all procedures. After further training with Naval Inspecting Officers they took over from RNVR officers.

Approval was given in June to recruit locally in Kilindini, Kenya and Washington, USA.

I was working with the British Routing Liaison Office in New York from 1941. Towards the end of 1942 we were asked to join the WRNS, which I did

Class 12 (L) in the summer of 1944.
(Lent by Mrs Wakem)

as a Writer. I later changed my category to Wren Coder. There was also a small band of Wrens at the Brooklyn Navy Yard and, of course, quite a large number in Washington.   Mrs J. Gallagher

In June a Telephone School opened at Westfield College, and a WRNS Writer School opened at Wesley College, Leeds.

The General Service Training Depot, Tulliechewan Castle, Balloch opened in June for Scottish and Northern entrants.

Fighter Direction Officers
In June WRNS officers were trained as Fighter

Ceremony of Colours.

The British War Relief Society of the USA, through
the Dudley House Committee, presented nine
wedding gowns to the WRNS. Officers and ratings could
borrow the gowns when, through lack of material
and coupons, they could not provide their own.

Her Majesty The Queen took the salute at the march past of 1,500 officers and ratings outside Buckingham Palace on 11 April 1943, the fourth anniversary of the re-inauguration of the WRNS. In attendance are (from left to right) the Second Sea Lord, Director WRNS, First Sea Lord, The Duchess of Kent and Mr A. V. Alexander.

Direction Officers. They took their turn as watchkeepers, checking all airborne aircraft within their area and homing lost aircraft. By December 1945 there were 36.

Mrs H. Gibling joined in 1941 as a Radar Plotter and was commissioned in December 1944. After a variety of courses she was appointed to Ronaldsway for Fighter Direction duties. One of her memories is as follows:

> At 2230 in the ADR we heard a plane's engine cut in take-off. . . crash buzzer went, the doc' flew out, you could see flares. Crash tender rushed over, but

the plane didn't catch fire. . . pilot had taken off with dive brakes down. . . whole crew escaped OK. Then a Barracuda landed plonk on the intersection of the runways with its undercart up. . . eight or ten planes were kept orbiting in circuit for one and half hours. Another Barracuda burst its tyre and held up all the traffic for half an hour. . . a shambles.

Air Synthetic Trainer

In the summer a new category was formed to include all Wrens working in Torpedo Attack Teachers, Link Trainers, Silloth Trainers and Operational Crew Trainers. It was always a small category, 35 being its peak number.

> In November 1943 I was drafted to Macrihanish, Argyll, to the Link Trainer Section. The other Wren with me was Kay Mallon. We were sent on a Link Trainer Course to the RAF Station at Hinstock, Shropshire. . . the only two women on the course, as I recall, and I am still proud to say that

The Wrens march past Her Majesty The Queen on their fourth birthday. After the parade there was a special service in Westminster Abbey.

we both passed whilst a couple of the men did not!

Mrs R. D. Matthews

Aircraft Engineering

Nine officers were trained for Aircraft Engineering duties. One was appointed for air maintenance, one as a Strip Camera Operator and a few as Link Trainer and Navigational Air Analysers.

I joined up from my home town of Torquay in June 1942 and worked with a Combined Operations Unit concerned with landing craft and engines. I was recommended for promotion to officer rank and was drafted to HMS *Heron*, Yeovilton, to gain

more knowledge of aircraft instruments, aircraft investigations, and attend some training sessions for Air Mechanics. I went to OTC Westfield College, and was appointed as Assistant Air Engineer Officer at the School of Aircraft Maintenance, Worthy Down where, among other duties, I was responsible for part of a conversion course for Senior Engineer Officers who were transferring from steam to air Engineering.

Mrs T. A. Leach

Boat Drivers

Training as Boat Drivers included the operation and maintenance of various types of petrol and diesel engined motor boats. 154 ratings were trained in all.

One afternoon in 1943 when I was coxswain of a small motor launch, I and my crew of one, Wren Kneebone, were on our way down river to pick up some 'chippies' from a trawler which was moored in the River Dart. We heard a noise which sounded like machine-gun fire. . . we felt the blast of a bomb . . . a large collier lying just astern was sinking with her crew floundering in the water. We hadn't much time to feel frightened but jumped into our launch and set out to help. . . everything and everybody was black with coal dust, and you could hardly distinguish men from driftwood. We worked furiously to help them aboard the launch. . . we shipped a good deal of water, and the engine got soaked and wouldn't start. Another small boat came alongside and took off the worst of the wounded. . . we had to abandon ship and climb aboard a minesweeper also in a pretty damaged condition. . . we could feel her sinking. After about ten minutes, two MTBs appeared – Barbara went in one, and I went in the other, each with over twenty survivors. My first duty on getting ashore was to report to the Commander that I had lost my boat. I won't repeat what he said. The boat was salvaged and it wasn't long before both we and our boat were once more clean and back on duty.   Patricia Konig

Boat's Crew in action.

I was the first Petty Officer Wren Coxswain of a harbour launch in Plymouth in 1944. I was chosen as the first woman to train as a pilot of HM Ships in addition to my normal duties. These ships were landing craft which, in some cases, were towing large barges. I led the craft through the channels from Plymouth Sound to the small village of Calstock, a distance of about eleven miles. My pay for this work was ten shillings per trip. After six trips the Admiralty said, 'No more money'.

Mrs P. Crossley

Teleprinter and Switchboard
In 1943 the Trafalgar Switching Centre for teleprinter work was taken over and staffed by WRNS personnel. The Post Office arranged training for maintenance and a small separate category was formed.

Chief Wren Travelling Supervisors were appointed in July 1943 to carry out the general supervision of naval telephone traffic and to monitor standards of operating. In the summer of 1944 there were 9 Supervisors, 1,926 Operators at home, and 120 overseas.

Wrens on Parade in Beirut – United Nations Day 1943. (Lent by Miss C. M. Baker)

WRNS officers were appointed to the Admiralty in July 1943 for duty in Intelligence, as personal assistants and secretaries and, in some instances, served as Flag Lieutenants.

Voluntary recruiting closed on 29 July with the introduction of the National Service Act.

In August a Chief Officer WRNS was appointed to the Staff of Commander-in-Chief Eastern Fleet and moved, with the first draft, from Kilindini to Colombo.

Censor Officers

In August nine WRNS Censor Officers were sent to Ceylon. Offices were opened at Bombay, Karachi, Calcutta, Vizagapatam, Madras and Cochin. All censorship personnel, with the exception of one Captain RN, were WRNS officers, with a Chief Officer as Chief Censor (Naval) at Colombo. The following is an extract from a report on naval censorship on the East Indies Station.

> The employment of WRNS officers on censorship duties appears to have been very satisfactory. A very high standard of discretion was maintained both on security matters and on personal details read in letters.

WRNS Censor Officers were employed in other Commands abroad and Mrs J. Buhaqiar was one of those in Egypt.

> After spending a year in the Fleet Mail Clearing Office at Ras el Tin, HMS *Nile* in Alexandria, I attended an OTC at Ras Rassafah and was sent off to Suez to be a Censor Officer at Navy House, Port Tewfik. Nothing could have prepared me for the shocks of the job. Sailors writing home to their wives and girlfriends do not understate matters – I had been gently nurtured and, at the age of twenty-two was highly romantic, and believed in keeping sex at bay! My happiest recollection of those duties came just after my marriage, in March 1945, to a

The Printing Office Staff, Alexandria, September 1944. (Lent by Mrs Drury)

Shipping Agent on the Canal. Absolutely all local naval personnel were invited to the wedding – we were a small, intensely friendly, group. So it came about that upon resuming my duties I found the usual batch of mail awaiting inspection. This extract was from a letter sent to his wife by a Leading Seaman:

> We were all invited to a wedding last week, and a slap-up do at the French Club after. One of our Wren officers married a local chap, and I tell you it was big eats and booze, as much as you like. She's not much to look at, but has a nice disposition.

Of course most of our duties were serious.

Night Exercise Attack Teacher

A few WRNS ratings were employed on these duties at Londonderry and were returned as a separate category the next year. There were never more than five.

WRNS Education

In September 1943 a WRNS Officer was appointed to the Rosyth Command for educational duties. This was later extended to each Command and a large number of WRNS Unit Education Officers were also appointed.

Aircraft Recognition

Four officers were eventually employed in this branch and their duties included secretarial, preparation of material and instruction on both aircraft and ship recognition.

Between 6 and 30 October an exhibition of Wrens' handicrafts was held at the National Gallery. It was opened by H.R.H. The Duchess of Kent and visited by Her Majesty The Queen.

The first WRNS Officers' Naval Control of Shipping course was held in October.

Twenty WRNS officers were aboard the troopship *Marnix van der Silt Aldegorde* in November when she was torpedoed in the Mediterranean. All the troops were rescued and put ashore in north Africa. A week later the WRNS officers were on their way to their original destination.

Air Radio Officer

By the end of 1943 there were nine WRNS Air Radio Officers who were responsible for the technical aspects of all radio equipment in aircraft, and on the ground. There were 39 by the end of 1945.

In December 1943 there were 3,888 Officers and 62,612 Ratings.

1944

Voluntary recruiting was re-opened for women under the age of 19 on 3 January.

By January WRNS officers were being admitted to the Staff College. Ten officers, in all, took the course. WRNS officers were being employed, increasingly, in staff duties. For instance, at ACHQ Plymouth in June 1944 they were in the following appointments: Assistant Staff Officer Convoys, Staff Officer Movements, Assistant Staff Officer Intelligence, Assistant Communications Officer, four Duty Signal Officers and five Plotting Officers.

In HMS *Ferret* and HMS *Philante* they also took over the duties of Exercise Officers, where they co-ordinated training programmes for the escort ships when in harbour.

The Welfare Staff, Belfast, at a wedding. As the war progressed the number of WRNS officers and ratings gradually increased in the very important task of assisting and advising naval dependants. (Lent by Mrs Donaghy)

Armament Stores Officers

There were 19 in all, selected from WRNS ratings Air Mechanic (O) and Qualified Ordnance categories. They were responsible to the Gunnery Officer for demanding all armament and explosive stores, and their issue, plus care and maintenance of equipment and explosive store houses.

Seventeen officers and 195 ratings deployed to Malta in February.

In March Portsmouth Command was given priority in drafting of new entries; and a small WRNS unit was established at Caserta, Italy. The next month the WRNS OTC had to move from Greenwich to Framewood Manor, Bucks because of bomb damage, and in May a new category of Coder (S) was

authorised for typex duties overseas – at peak there were 106.

In May the shortage of Motor Transport Drivers was so acute that a loan was arranged of 150 WAAF drivers, for a period of three months, until in-service training started in August.

In June a Superintendent WRNS, Mediterranean was appointed to the Staff of Commander-in-Chief Mediterranean. The Levant then became an area under a Chief Officer WRNS and included Alexandria, Cairo, Port Said, Ismailia, Suez, Haifa and Beirut.

I joined the WRNS in 1940, served as a Writer at HMS *Daedalus*, as a Third Officer at HMS *Raven*, *Daedalus* again and then to Alexandria. In September 1943 I was posted to Haifa which was then the good ship *Moreta*. . . Haifa was the ship repair base for the eastern Mediterranean. . . At the time I was sent to Haifa there were only four WRNS Officers (Cypher) there. This was gradually increased to

Four Petty Officer Wrens relaxing at HMS Assegai. (Lent by Mrs Baldwin)

ten, including a Fleet Mail Officer and a Censor Officer, as well as extra Cypher staff. Later still a number of Wren ratings came, which meant another mess and the appointment of a Quarters Officer. **Mrs J. Colenutt**

D-Day

On 6 June 1944 a large force of British, American and Canadian troops landed on France's Normandy coast. Everyone who was in the WRNS at this time has her own memories of the events surrounding this most important date. Below are just a few comments and anecdotes of before and after D-Day.

Wren Motor Transport Drivers were responsible for the running maintenance and cleanliness of their vehicles, as well as appearing in pristine elegance as chauffeurs.

WRNS Boat's Crew served the British and American Fleets, delivering stores, despatches, signals and ferrying Commanding Officers ashore for sailing orders. Some WRNS Coxswains were trained to serve as pilots for D-Day, taking the smaller ships on their way and drawing pilotage money.

Shortly after D-Day I was ordered to take two boats out into the Solent to meet a disabled LCI(S), and tow her into the Hamble River. When we found the ship she was already being towed by a sister ship, but another LCI(S), No. 353, was down at Gosport sinking – could we do something about her? She was lying with her bow completely under the water, and propellers and rudders high in the air. My other boat arrived, so we lashed one boat on each side of 353 as far aft as possible, and started to tow her stern first. . . I took my stand on the highest part where I could be seen by both boats

APRIL 1944   No. I.

Levant W.R.N.S. Review

JOINT EDITORS: SECOND OFFICER M. FLETCHER.
THIRD OFFICER A. WELLS.

EDITORIAL

As this is the first issue of the Levant Review, it has to cover a period of two years. The first draft of wrens to the Middle East arrived on March 8th, 1942, and they have been coming in quick succession ever since. Alexandria was the first home of the wrens in Egypt, but during the past two years quarters have been opened in most of the ports in the Levant and Middle East Area.

The present editors, who took over during the production of this issue from Third Officer Collingwood, would like to make it known that it is hoped, with adequate co-operation, to speed up production of the Review, and make it a quarterly magazine. Please help to lighten the burden by sending in contributions (especially items of interest to people in England), signed or anonymous, frivolous or serious.

We should also like to send our very sincere thanks to Third Officer Collingwood for her admirable work in connection with the Review, and for collecting and arranging nearly all the material for this issue.

3

Above: *Second Officer M. Fletcher and Third Officer A. Wells jointly edited this review which included informative articles as well as births, marriages and deaths. (Lent by Mrs Ambrose)*

Opposite: *A mixed naval congregation somewhere in England.*

Ship Ahoy!

and, by our own code of visual signals, I could control the speed of both boats. . . after some three hours towing we got within sight of the base and I asked the CO to call up the signal station and ask to have the jetty cleared for us, and the big fire-pumps standing by. . . when we got into the river it seemed to us that the whole ship's company, from the Captain down, was waiting for us. . . when we put her on the hard at high water, we could see what was wrong. She had a hole 11 feet long in the bottom under the troop compartment, received when she ran into an underwater stake on the Normandy beaches.

Three days later, 353 was off once more to France, and our reward was her CO's remark that he did not care what happened to him in the future; whatever it was the WRNS would see him through!

(Extract from an article in October 1961 issue of *The Wren*, by **Lita C. Edwards**)

WRNS Writers were responsible for the vast amount of clerical work in connection with orders and plans for the invasion of Europe.

Mail Clerks, Censor and Fleet Mail Officers handled the mail for the thousands of men ashore and afloat. Total censorship was imposed a few days before the invasion. Eighteen new Port Censorship Officers were established and the 26 WRNS officers were reinforced until there were over 200 officers carrying out these duties. It is estimated that between 25 May and 5 June WRNS officers at the ports censored 400,000 letters. The Chief Naval Adviser (Censorship) reported:

The general security of the mail was excellent and it was quite evident that both officers and ratings fully realised the need for discretion. . . the work which the individual censor is required to do is exacting, often monotonous and never popular, but civilians and WRNS officers alike have worked consistently with both enthusiasm and intelligence.

WRNS Supply Assistants at all Combined Operations bases undertook a large part of the issuing of clothing, victualling and naval stores. Between April and June 1944 the small team attached to Force J dealt with all aspects of stores for over 100 ships a day.

WRNS Cooks and Stewards before, and after D-Day provided for Portsmouth Combined HQ at Fort Southwick two wardroom messes above ground, and one below. Approximately 800 meals were served daily and a 24-hour service maintained; they dealt with a great influx of officers and managed a wide variety of permanent and temporary accommodation. In the WRNS mess roughly 1200 meals were served every day, including underground. In the latter, meals were served in five shifts, because of lack of space, and the Wrens were cooking in temperatures of up to 100 degrees Fahrenheit.

'Hello Jenny' – Wren Cooks on the way to the field kitchen – a very necessary art in the event of an invasion.

I was a Leading Wren Cook. I remember the day I arrived in Eastbourne; bombs dropped on Marks and Spencer's and an air raid shelter was demolished, killing sailors and Wrens. Most of the civilian population of Eastbourne were evacuated. . . Some weeks before D-Day Eastbourne came to life with the Army, mostly Americans, moving into all the empty property. . . On 6 June we could hear them moving out all through the night, and the next morning everything had gone and Eastbourne was a ghost town again except for the RN, WRNS, RAF and WAAF.   Mrs V. C. Goodall

I went to HMS *Northney II* shortly before D-Day where I worked for the officers at a house called Lennox Lodge. . . back down on the old knees once more polishing halls, bathrooms, cabins. Another Wren and myself were ordered to scrub out prefab holiday bungalows on one end of Hayling, part of HMS *Dragonfly*. . . they were empty and the floors were filthy. . . only one had water turned on so we

had to keep filling our heavy metal buckets with cold water, no hot, and armed with hard deck scrubbing brushes and yellow soap blocks, we set to. Thank goodness it was summer. . . I suppose we were helping the war effort in our way.

Mrs N. M. Hunt

Communicators were in the thick of it, but Mrs E. M. Foxon's memory is of the ships and boats.

I was a Communications Wren based at HMS *Osborne* on the Isle of Wight, working in an old fortress, Fort Bembridge. At this station we were concerned mainly with air sea rescue boats, but our work load was greatly increased at the time of the Normandy landings. . . Thousands of ships and boats of all shapes and sizes left from the Solent, and on the evening before the landings we felt we could have walked across to the mainland by stepping from one ship to another.

Switchboard Operators were particularly busy as planning for D-Day gathered momentum. In the October 1960 issue of *The Wren* Jean McCormack, who had worked on the Portsmouth Combined Headquarters switchboard at Fort Southwick since June 1942, recalled what it was like during this very important period.

Shortly before D-Day one visitor described the main PBX: 'as brilliantly lit as Piccadilly Circus before the war'. Traffic on all boards was exceptionally heavy. In addition we had to open a separate enquiries section which worked at pressure to get records up-to-date. . . no easy task when 100 new circuits were allocated in one day. We were also working out procedure, and training supervisors and operators to use the Radio Monitoring Units (now known as radio telephones). The great day drew near and tension mounted hourly. Calls on the switchboards were drastically restricted. . . At last D-Day arrived! Traffic gradually slowed down until everyone's interest was centred on the plot. Then came the great moment. . . we had our first call from the far shore! What joy to hear the voices of the men who had operated radio units with us, in and near HQ, and were now putting our procedure into practice. . . My most vivid personal recollection is of taking down details of the bombardment of Cherbourg – wonderful to

feel you were in personal contact with all ships concerned, but terribly sad to report direct hits almost as they were made.'

During June, mainland Britain was facing a new threat from the V1 flying bombs, though at first their effect seemed to be one of instilling curiosity rather than fear.

As a Wren Writer working in Chatham Barracks and quartered at Gadshill in June 1944, I believe that I may have been one of the first to see a V1, the flying bomb which soon became known as the 'doodlebug'. On this night in early June I stopped to exchange a few words with the two ARP Wardens and, as I did so, what appeared to be a plane in flames flew at eye level parallel to the road and crashed into a nearby field. I said 'Poor pilot', but the Warden answered, 'We don't think there is a pilot in that plane'. Next day, in the Commodore's office where I worked, we were told that the 'plane in flames' was a flying bomb. I believe that this may have been one of the ten V1s launched on 12 June, of which only four arrived. In the days that followed the low droning of these 'doodlebugs' became a familiar sound, as did the sight of fighters chasing and trying to shoot them down.

Mrs C. Coulthard

At Western Approaches Tactical Unit, Liverpool the Captain of the WATU noted the following.

Those who have seen this team at work will understand that the WRNS officers and ratings have neatly stepped into the shoes of the Naval Staff. . . In a tactical game the Commanding Officer of each ship must have one of the staff to point out the situation in the game, as it affects their Command and those adjacent, and to obtain the officer's orders for course, speed, gunfire, asdic attack or smoke screening, every move of the game. This must all be done in seamanlike language. . . In these exercise actions a Wren Mover may be asked to explain the best action to be taken in order to compete with any situation. The trust placed in the advice given them is uncanny. This technical and tactical knowledge is the result of the training of WRNS officers and ratings by the Naval Staff of the Tactical Unit, supplemented by Courses, flights in

Fleet Air Arm aircraft, and visits to Escort Vessels during trials and exercises.

A light-hearted memory of June 1944 was provided by L. N. Trye, an ex-Third Officer WRNS in the October 1961 issue of *The Wren*.

A call for volunteers to help lift the potato crop came from the local farmers. . . Our party consisted

The WRNS unit, HMS *Tana, Kilindini, Mombasa; Christmas 1942. (Lent by Mrs Vane Percy)*

of eleven Wrens and WRNS officers, two RNVR officers and four Canadian naval officers – most of us attired in slacks and jerseys. We worked in pairs, with a bucket and sack between us. A pair of horses was harnessed to a mechanical plough with a revolving fork which flung the potato plants to one side and stripped off most of the potatoes at the same time. These were then gathered by hand. . . Time passed very quickly and at ten o'clock we called it a day.

A final note on this period shows something of the affection the Navy had for their Wrens. The following extract is from a letter from HMS *Borage* to the W/T Station at HMS *Nimrod*, Campbeltown where Mrs J. Hatfield was stationed:

What would I not give for another quiet month at your place, each had agreed that it was the most pleasant of times since joining the Navy. What bricks you all were to make our stay so jolly and it has provided a memory that is a milestone in this irksome journey to the end of the war.

Unanimously agreed that Campbeltown has the nicest and most charming W/T staff in the Navy.

In June a First Officer WRNS was appointed to the Staff of Flag Officer Air (East Indies). Local climatic conditions and customs precluded some categories, but SDO Watchkeepers and Coders were employed in large numbers, as were Writers and supply ratings. Twenty-three Topographical ratings were based at the SEAC Inter-Services Department, and some air categories were also employed in the Command.

North-West Europe

Plans for the employment of WRNS personnel were in hand before D-Day. In July 1944 the Deputy Director (Welfare) and a Chief Officer went to France to make arrangements. On 15 August the first party of WRNS personnel arrived in Normandy, for duty with Flag Officer British Assault Area at Courselles, consisting of two officers, one Petty Officer Regulator, one Petty Officer Quarters Assistant, three Cooks and four Stewards. These were later joined by Cypher and Secretarial Officers, and Writer and Communications Ratings (14 officers; 84 ratings). A unit of 11 officers and 42 ratings was later established with

NOIC Arromanches. At the end of August, 45 officers and 146 ratings transferred from ANCXF, Portsmouth to Granville, Normandy. The Courselles unit moved to Rouen in September, the Arromanches unit moved with the HQ to St Germain-en-Laye and, by January 1945, units were formed in Brussels and Ostend.

After a year in Signals on the Isle of Wight I returned for a course in London to qualify as a Teleprinter Instructor. By this time preparations were well under way for D-Day and I returned to the south coast and was drafted to Southwick House, the Portsmouth HQ for the D-Day planners. Working in a converted wine cellar whilst, just above our heads in the now famous library, Generals Montgomery and Eisenhower and Admiral Ramsay were often present finalising some of the last details of the big day, and the subsequent assault on Europe.

Following the successful landings in Normandy I was among the first WRNS personnel who, in August, embarked in an LCT to sail for Normandy as part of a Naval Signals Unit. Our party disembarked on the Mulberry harbours at Arromanches and then went across country to our first base at Granville, until it was safe for us to proceed to the HQ in Paris... As our troops pushed forward, some three months later, we moved on up to Brussels.   Mrs M. I. Preston

Mediterranean
In July 1944 the Superintendent WRNS Mediterranean moved, with the Commander-in-Chief, from Algiers to Caserta where a few WRNS officers and ratings had been employed since March 1944.

The Superintendent was responsible for Gibraltar, Algiers and Malta, whilst a Chief Officer in the Levant was responsible for all Middle East units. The majority of WRNS personnel were Writers and Communicators. Large numbers of Pay Writers were required in Malta and Alexandria; some Cooks, a Mess caterer and a Steward were employed in Gibraltar but, apart from the Quarters Assistant, all domestic labour was recruited locally.

One Wren Motor Transport Driver was in Gibraltar and, apart from a not very successful trial of 12 Drivers in Durban, because of local difficulties and customs, few drivers were employed abroad.

HMS *Aurora* transferred Wrens from Algiers to Naples in August. An excerpt from Algiers' *Union Jack* reads:

A sensation was caused in Naples harbour recently when HM Cruiser *Aurora* arrived from Algiers, for not a man was to be seen on her forecastle. The only man in sight forward of the bridge was the First Lieutenant who was at his usual station.

The *Aurora* had been the only ship available at the time to take forty Wrens from Algiers to Naples for service in Italy... During the voyage they were assigned duties about the ship... for the first time in the history of the Navy, one of His Majesty's Cruisers entered harbour with Wrens, in their formal white dresses, manning the fo'c'sle.

As she proceeded to her anchorage, the *Aurora* had to weave in and out between other ships. At first her approach was viewed by ratings of nearby ships with normal interest, which rapidly changed to astonished incredulity as the onlookers realised that the vision was a reality.

At home the Service continued to grow for some months after D-Day until the peak number of 74,620 was reached in September 1944. After that the numbers fell and fluctuated until June 1945, when releases began under the Government's Release and Reallocation Scheme.

The first three WRNS officers were appointed for a Flying Control course in October. Five officers served in this branch.

WRNS personnel were much involved in amateur entertainment throughout the War. Mrs C. M. Lowry recalls the following.

In conjunction with ENSA a naval concert party was arranged to tour France to entertain the troops. The company, 10 Wrens led by Third Officer Hazel Wilkinson, and 10 naval personnel, led by Lieutenant Eric Barker RNVR assembled in Dover in December 1944. They waited for two days for the minesweeper HMS *Foulness* to take them off, and were issued with Arctic clothing for the four and a half hour Spartan voyage. They toured the French and Belgian ports for four weeks, entertaining all Allied servicemen who were waiting to move into Germany.

In December 1944 there were 4,646 officers and 69,309 ratings.

T O *Betty Charlotte Hopkins*

You are hereby appointed as

Third Officer

of the

WOMEN'S ROYAL NAVAL SERVICE

and are requested to enter upon your duties

in *H.M.S. Pembroke III*

and to observe and execute the Instructions
for the Government of the Women's Royal Naval
Service and all such orders and instructions
as you shall from time to time receive from
your Superior Officers.

Vera Laughton Mathews

DIRECTOR W.R.N.S.

3. 7. 44.

An example of a wartime letter of appointment.
(Lent by Miss M. R. Hopkins)

1945

Wrens were still being drafted abroad and, in January, Mrs J. Shea embarked for Aden.

We finally came to a halt on the quayside of the KGV dock in the Clyde. Here there were hundreds of troops waiting to board the ss *Almanzora*. My cabin was quite large, fortunately, as there were thirty of us squeezed into three-tier bunks. Portholes sealed, and everything dark and hot – each night the Duty Officer would make spot checks to ensure that Wrens were wearing bell bottoms – anyone caught in pyjamas would be in serious trouble. . . Water was turned on for half an hour morning and evening which meant the 'Charge of the Light Brigade'.

One of the most treasured memories of all my time took place in the middle of a storm. Sunday morning, six o'clock ship's time, making my way through the darkened ship to an appointed place. There I found some ten to twelve people, indistinguishable in the dark except for the glow on each face from the lighted candle each was holding – and so we knelt, candle in hand, on the bare boards and took our Communion.

The Maintenance Unit, HMS Jackdaw, *Crail (Lent by Mrs Farrell)*

As the war in Europe drew to a close WRNS categories were reviewed to identify those which would become redundant, continue, or be enhanced.

Releases began in March, married personnel being in the high priority groups.

In May Civil Servants were recruited into the WRNS for clerical duties in dockyards at Colombo

and Trincomalee. Shortfalls were met by re-mustering Wrens from redundant categories. The numbers required were never achieved and the first draft did not leave the UK until 1946.

Welfare Workers

Since April 1944 certain members of the Administrative category had been employed as temporary home helps for naval families in cases of sickness and emergency. In May 1945 these ratings became Welfare Workers. After training they were advanced to Chief Wren and were employed in six Commands. In December 1945 there were 54.

The Air Command was created in June 1945 and a Superintendent WRNS was appointed to the Staff of Admiral (Air).

The first WRNS return showed 14,638 personnel were working with the Fleet Air Arm.

EVT

Training started for WRNS EVT Instructors in June, and a new category was formed in December. They taught academic and craft subjects. Local instruction was already in being and Mrs M. E. Andrew was one of those who benefited from this facility.

During the War, whilst I was stationed at HMS *Haig*, Dover, I received permission to marry a Canadian soldier. I went home on leave to get married, but my fiancé was in the middle of the Mediterranean

Two Wrens and a Nurse in the Kurfurstendam, October 1945. (Lent by Mrs Curry)

picking up survivors from a torpedoed ship. . . In 1944, it was decided that watchkeeping Wrens should have sewing classes to keep them occupied after so much time underground. I asked if I could make my wedding dress. A signal went to the Admiralty requesting permission to purchase six yards of bridal satin; we were quite surprised when permission was granted. Our instructor, a PO Wren, I believe, was most proficient, and helped me make my beautiful wedding dress, entirely by hand. My fiancé came back from Europe after eighteen months and we were married in 1945.

In June the WRNS OTC returned to Royal Naval College, Greenwich from Framewood Manor.

In July a small party of specially selected WRNS ratings were sent to the Staff of Flag Officer Malaya, at Kurunegala. One of the ratings typed the terms of the Japanese surrender.

ANCXF unit moved from St Germain to Minden in July, and the next month the Brussels unit moved to Hamburg. Eventually WRNS personnel served in Kiel, Wilhelmshaven, Cuxhaven, Bremen, Hanover and Berlin.

Following a month in Minden I was drafted to HMS *Princess Irene*, Berlin. Another Wren and I were sent off in an army vehicle with a young Marine driver. It was a long journey and, after Hanover, we had to make for the one place where we entered the Russian Sector before getting to Berlin. We only had our flimsy draft papers – it wasn't surprising that we got lost and were turned back, eventually finding ourselves nearer to Hamburg than anywhere. We naturally thought Hamburg would inform Berlin of our whereabouts, but apparently not – we learned later that the WRNS officer thought that 'a fate worse than death' had befallen us. . . she had got the entire Naval establishment out looking for us. We dined out on that story for a good few weeks after our arrival.　　Mrs J. Curry

By July AA Target Operator, Boom Defence, Courier, Special Duty (Linguist) and Coder (S) had been declared redundant. At the same time the following categories were to receive no further recruits:

Air Mechanic (A)(E) (L) and (O)	Automatic Morse Transcriber
Anti-Gas	Boat Driver

*Demobilization party in Berlin in 1946.
(Lent by Mrs Curry)*

We sailed into Sydney with colours flying and bands playing on the quayside – we were so thick on the port side that they couldn't tie up owing to the list, and we had to be shooed away. The *Orontes* was the first of her line to call there for four years. . . I was with the Melbourne draft and, after a 21 hour journey aboard ancient and dilapidated troop trains we arrived. In August we were all feverishly buying and despatching food parcels for home – everyone was full of the Japanese offer to surrender, and Melbourne went mad that day.

Miss H. L. Andrus

Mrs E. Graves' journey to Sydney in December 1944 was subject to considerable delay and, instead of arriving two or three weeks before the main party, they arrived the day before. There were fifteen in the party which sailed in the *Queen Elizabeth* to a New York blazing with lights, then on to San Francisco by train to a warm welcome and Christmas with friendly families. Next by Admiral Nimitz's private plane to Honolulu and seaplane to the New Hebrides. Here, unfortunately, German measles was diagnosed, and because of the susceptibility of the local inhabitants to the illness they were quarantined for two weeks. They eventually arrived in Sydney to start up the communications office.

Back home Mrs M. J. Carter recalls:

My most memorable day, and the one I still remember vividly, occurred during the night of August 14/15. I was lying in my bunk, but still

Boat's Crew	Plotter
Degaussing Recorder	Printing
Gunnery Control	Qualified Ordnance
Gunnery Experimental	Ship Mechanic (LC)
Assistant	Special Operator
Maintenance	Torpedo Wren
Maintenance (Q)	Visual Signaller
Net Defence	Wireless Telegraphist

Parachute Packers and Safety Equipment Workers were incorporated into a new category called Safety Equipment Assistants.

By August 102 officers and 442 ratings were in Australia at Melbourne, Sydney, Brisbane and Herne Bay.

Convoy 'D' en route from Melbourne to Sydney in December 1945. (Lent by Miss Andrus)

Wrens in the Victory in the Pacific Parade in Melbourne. (Lent by Miss Andrus)

awake, when at 0001 I heard a ship hooting its siren in the harbour. Immediately afterwards all the other ships took up the sound. Everybody was awake by then and we turned on the radio just in time to hear the wonderful news that Japan had surrendered. We all leapt out of our bunks and stood outside our hut and listened to the sound of all the ships' sirens, and then watched fireworks and searchlights which lit up the pitch-black darkness of the black-out. . . after a while, the Wrens in our hut donned bell-bottoms and sou'westers, as it was pouring with rain, and strolled around the camp watching the various bonfires which the sailors of HMS *Impregnable* were lighting. . . we were all back in bed by 2 a.m. – but nobody slept for a long time.

In August some Wrens, who had worked with British Naval Liaison to the Royal Norwegian Navy, paid a courtesy visit to Norway.

I worked in Isolated Units from 1942 to 1946 with the Royal Norwegian Navy in London. PO Wren Dorothy Wilson and I were both offered a courtesy visit to Oslo immediately after the War, as a token of their appreciation. We had to go singly, as one of us had to stay behind to mind the ship. During my visit to Oslo I was invited to the palace. Many wonderful and brave Norwegians became my friends. The anecdote which stays in my memory of my visit is, whilst waiting for a bus, a German jeepload of British sailors – a lovely sight – spotted me and yelled out, 'The Wrens are here!' Little did they know I was the only one and the first, incidentally, to set foot on Norwegian soil.

Mrs E. N. Bliss

*MT Section, WRNS, London Transport
Instruction Centre, Chiswick, July 1945.
(Lent by Mrs Shaw-Reynolds)*

WRNS personnel serving overseas were to be involved in the administration of prisoners of war returning home.

I boarded *Athlone Castle* and sailed for Ceylon on 2 October where I was seconded to Government House to sort Far East prisoner of war mail recovered by the Red Cross. Our job was to check names of POWs returning on troopships to the UK, against piles of mail. It was a sad task with the quantity untraceable far exceeding those returning home.   **Mrs M. O. Somers**

The following passage from a letter written by Chief Officer A. McNeil, in the Levant, to the Director WRNS, was re-printed in the January 1946 issue of *The Wren*.

At the moment we have a party of Wrens on loan to Suez helping with the reception of returning Prisoners-of-War. . . All the Wrens are volunteers and doing an excellent job. Two ships came in the

day I was there. Most of the people on board looked much better than one had imagined, but I think it is probably the fitter who are coming through now, and of course they have had over six weeks' feeding up. . . the whole object of the ships calling at Suez is for all returning personnel, whether service or civilian, to be given suitable clothing for the kind of weather England is likely to produce on their arrival!

In October WRNS personnel moved from Kandy to Singapore as part of Admiral Mountbatten's Staff, and Wrens from the UK were drafted to Hong Kong.

In the Autumn Wrens took part in an Allied naval exhibition in Rotterdam and Amsterdam.

On 1 December 24 officers and 9 ratings transferred from Sydney to Hong Kong to the new headquarters of Commander-in-Chief, British Pacific Fleet.

At the Annual Dinner of the RNVR (Auxiliary Patrol) Club the Toast was the 'Women's Royal Naval Service'.

In the presence of the First Sea Lord, the Lord Mayor of London, and several hundred members of the Club and visitors, Vice Admiral Sir Denis Boyd paid

Visit by Director WRNS to Colombo, November 1945. (Lent by Dame Nancy Robertson)

tribute to the WRNS. He referred to the early days when, as Captain of HMS *Vernon*, he had been one of the first Commanding Officers of Wrens. He said:

About one hundred arrived in plain clothes, with one WRNS officer and, at first, I felt very much as the Duke of Wellington did on getting some new recruits during the Peninsula War. He is reputed to have said, 'I don't know whether they will frighten the French but, by God, they frighten me'. Indeed our Wrens did frighten us, but it was interesting to see how they broke down that inhibition. They astonished us then and still do. The Navy loves its Wrens, and life ashore at the depots, the airfields and the offices has been made bearable by the brightness and enthusiasm of these girls, a huge body of women welded together into an efficient force and with a spirit which is unequalled. Not a force of Amazons but women who, however good at their work, have remained women and an influence

for good on all those who have had the privilege of working with them.

In December 1945 there were 2,879 Officers and 45,483 Ratings.

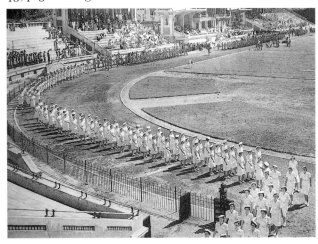

Thanksgiving Sunday Victory Parade, 13 May 1945, in The Stadium, Alexandria. (Lent by Mrs Currie)

CHAPTER 4
POST-WAR TO 1949

The post-war years of the Forties were to be a period of retrenchment and reorganization. The immediate requirement was to demobilise and return the majority of Service personnel to civilian life. However, in order to do this, a well conceived Release Plan was put into train which needed a large number of Wrens to administer it. Each of the home ports had its own organization to sort out pay, release documents, civilian clothing etc., and Resettlement Information Officers provided advice on careers.

There were, amongst some people, thoughts that,

WRNS Contingent in the Victory Parade in London. (Lent by Mrs Broster)

as at the end of the First World War, the WRNS would be disbanded. However times had changed and, with these changes, women's perspective of their working life had altered. Also the women were, by this time, an integral part of their parent Service having taken over, during the last five years, the major share, in the case of the Royal Navy, of the shore support functions. The retention of the WRNS was announced in a Government White Paper on the future of the Services as a whole, issued on 30 May 1946. By June 1946 the WRNS numbers were reduced to 15,000 and would continue to decline for some time to come. However on 1 February 1949 the WRNS became a Permanent Service, separate from but

WRNS Contingent in Scotland's Victory Parade in Glasgow. (Lent by Mrs Broster)

WRNS Central Training Depot for all new entries to the Service. It was called HMTE *Dauntless.*

integral with the Royal Navy, and a new era had begun.

In the summer of 1945 a former Ministry of Supply camp, at Burghfield near Reading, was taken over as a WRNS Training Depot.

It was used to accommodate the remustered ratings and all overseas drafts, thus allowing Crosby Hall, Chelsea and Westfield College to be de-requisitioned. Mill Hill continued to provide general service training until February 1946 when it was de-requisitioned and, from then on, Burghfield Camp became the

Mrs J. Newton was one of the last Wrens to leave Coimbatore, India in 1946.

We travelled down to Colombo to wait for a ship back to the UK. Whilst there I worked in the Fleet Mail Office. I finally sailed to the UK on board the aircraft carrier, HMS *Indomitable* and, as my father, brother and uncle all served in carriers I felt I was upholding the family tradition.

On 8 June 1946 a contingent of some 200 Wrens took part in the Victory Parade through London. The three Women's Service Directors at the end of the war – Dame Leslie Whately, Lady Welsh and Dame Vera Laughton Mathews – were seated in the Royal Stand.

On 4 July 1946 the inaugural meeting of the WRNS Amenities Fund Committee was held.

On 24 November a WRNS memorial window was dedicated in St Andrew's Church, Rosyth dockyard.

———————

Dame Vera Laughton Mathews DBE was succeeded as Director WRNS by Superintendent Jocelyn Woollcombe, in November 1946.

The following letter was sent to Dame Vera, by the Board of Admiralty, on 25 November 1946.

I am commanded by My Lords Commissioners of the Admiralty to convey to you, on the occasion of your relief as Director of the Women's Royal Naval Service, an expression of their high appreciation of the services which you have rendered during the tenure of that appointment.

On the 11th April 1939 you kindly accepted an invitation from the Admiralty to undertake the task of organising the Women's Royal Naval Service on a voluntary peace-time basis. Before the preparations for the new Service were completed it

The WRNS window in St Andrew's Church, Rosyth.

became necessary to bring it into active service when war was declared on the 3rd September 1939. The rapidity and success with which this was accomplished was proof of your enthusiasm, strength of purpose and organising ability.

You were successful in bringing into the present Service the experience and traditions of the previous Corps. Throughout its development to its highest strength of 75,000 you were able to infuse into it the high standards of the Royal Navy and to avoid many of those difficulties apt to beset a new Service. The wide range of activities in which the Women's Royal Naval Service has been employed is due largely to your inspiration and guidance.

In your responsibility for the morale and the well-being of the Service, you have won for the Women's Royal Naval Service, the trust and esteem of the public.

It was a source of great satisfaction to My Lords that His Majesty the King honoured you by appointing you to be a Dame Commander of the British Empire in the New Year's Honours List 1945.

My Lords are glad that you have been able to remain as Director of the Women's Royal Naval Service until the war-time purpose of the W.R.N.S. has been fulfilled and the process of transformation to the future organisation far advanced, so that the peace-time force will be built up on foundations laid by you. They greatly regret that the time has now come for you to relinquish the post of Director of the Women's Royal Naval Service and they desire me to convey to you their best wishes for your well-being in the future.

Three large vans fitted with complete kitchens and modern household appliances were to travel to many naval establishments and provide housecraft and dressmaking courses for WRNS personnel. They were operated by members of the WRNS Education category. The following item appeared in the April 1947 issue of *The Wren*.

One Saturday morning in March a large van moved out of London. It was the WRNS Education's latest venture commencing its long, five day journey to the east of Scotland. Two days later it was followed by a second van going to Cheshire; and in another two days the last of this trio set out for South Wales. . . Since then the vans have visited

Inside one of the household vans.

many stations, and have provided housecraft courses for more than 250 Wrens, and dressmaking courses for 50. Their journeys have taken them to Easthaven, Rattray and Arbroath in Scotland, to Stretton, Hinstock and Warrington in the Midlands, and further south to Dale, St Merryn, Devonport and Worthy Down.

Also in the April 1947 issue of *The Wren*, the Director WRNS wrote.

The future has seemed shrouded in a fog of uncertainty and even the assurance that the Women's Services are to form a permanent part of the Armed Forces of the Crown has not yet greatly illuminated the way, for we cannot yet see plainly the shape of our permanent force... The Navy must always be ready as a fighting force; we too must train in peace-time so that we may take our places in partnership should the need to defend our country ever arise again. . . We may never accompany Prime Ministers on vital missions; we may never plot the operations of a great invasion – indeed I hope we never shall – but will it not be well worth while if we can tell our grand-children that we played our part in establishing the first peace-time permanent Women's Royal Naval Service.

Dame Vera Laughton Mathews, in her last meeting of the Association of Wrens as Director WRNS, addressed the meeting as follows.

Obviously the sense of urgency cannot be so acute as when the safety of the country and all it stood for

depended primarily on sea power. But sea power must always be of vital importance to this country . . . I think it is wonderful that at last this ancient Service which is the very heart and tradition of Britain should have opened its doors to women and they should always feel that their work is of the utmost importance. We know that the decision to keep the WRNS as a permanent Service was due very largely to the manpower situation in the country, but that will not be so for ever, and the future of the WRNS depends very much on the account that those now serving give of themselves. They must make themselves essential to the Navy.

WRNS Central Sports Committee was brought into line with the Royal Naval Sports Control Board in July 1947.

Article in the January 1948 Issue of *The Wren*

It is now over two and a half years since the first members of the Service said goodbye to uniform and embarked on a peace-time career.

During that time there have been many changes – Wrens have come and gone, and the whole shape of the Service has altered considerably. The war-time figure of approximately 75,000 has fallen to 7,000 and, although the Commands are the same in structure, the shore establishments that compose them are now few and far between. The following lists will give some idea of where Wrens are serving today:

Air Command Lee-on-Solent; HMS *Ariel*; RN Air Stations Abbotsinch, Anthorn, Arbroath, Bramcote, Culdrose, Culham, Dale, Kete, Donibristle, Eglinton, Evanton, St Merryn, Worthy Down, Yeovilton.

Nore Royal Naval and Royal Marine Barracks, Chatham; HM Ships *Cabbala, Ganges, Ceres, Wildfire*; RN College Greenwich; WRNS Central Training Depot, Burghfield; London; Germany.

Portsmouth Royal Naval and Royal Marine Barracks; HM Ships *Excellent, Mercury, Royal Arthur, Osprey, Collingwood*; RN Auxiliary Hospital, Sherborne.

Plymouth Royal Naval and Royal Marine Barracks; HM Ships *Raleigh, Scotia*.

Rosyth Pitreavie; HM Ships *Lochinvar, Caledonia, Roseneath, Bruce, Sea Eagle*.

Mediterranean – Malta.

All drafting and advancement (with the excep-

tion of Air categories) is now carried out from the Training and Drafting Depot, Burghfield, near Reading; the successor for new entries of Mill Hill and Westfield.

With a smaller Service there are inevitably fewer categories and many of the exciting wartime jobs are no longer necessary. The Supply and Secretariat Branch employs the largest proportion of Wrens, and there has been a recent expansion in the Air categories with the training of Naval

WRNS personnel march past Her Majesty The Queen. (Lent by Mrs Broster)

Airwomen (similar to Air Mechanics (A)(E)(L) and (O)).

In the Communications Branch the W/T category has been re-introduced, with an extended syllabus and the title of Telegraphist. S.D.O. Watchkeepers and T/P Operators are now combined as Wrens Signal, and there are still a limited number of Switchboard Operators.

The question which is very much in the mind of serving Wrens is 'What will the permanent Service be like?' Unfortunately, at present, no one can say what the actual conditions will be or when they will come into force. It will be appreciated that there must be careful consideration of all the different angles before a permanent Women's Service comes into being, but it is hoped that when the announcement is made, those who wish to make the Navy their career will not be disappointed.

In the meantime the Women's Naval Reserve has been launched, and it is hoped that many ex-WRNS officers and ratings will decide to join.

Wrens were deployed to HMS *Falcon*, Royal Naval Air Station Halfar, Malta in January 1948. Recruiting opened for ex-Wrens, and others interested in nursing, to serve as VAD members with the Royal Navy on a two year contract.

There had been VADs during the war and Mrs J. Oxtoby was one of them.

Being in a reserved occupation I was only able to join up for the last two years of the War. The WRNS being full at that time I became a naval VAD and served at Gosport, Whitley Bay, Seaforth and Londonderry... Princess Marina, the Duchess of Kent, visited the sick bay at Whitley Bay and I remember well all the polishing and painting beforehand. The patients were issued with new pyjama tops, half an hour before the Duchess' arrival, which were promptly replaced immediately afterwards.

In the Spring of 1948 the Admiralty initiated a Women's Royal Naval Reserve, open to ex-officers and ex-ratings of the WRNS who had served more than twelve months, and who were prepared to undertake serving in the WRNS in the event of a future national emergency.

For the first time a Royal standard was broken over an all-WRNS establishment when Her Majesty The Queen visited HMTE *Dauntless* on 20 April 1948.

Above: *Her Majesty The Queen talking to Wren Cooks during her visit to* HMTE Dauntless. *(Lent by Dame Nancy Robertson)*

Below: *H.R.H. The Duchess of Kent takes the salute during a visit to* HMTE Dauntless *in March 1949. (Lent by Miss E. H. Scott)*

On 26 October the first Wrens to serve in the Middle East, since the end of the war, joined HMS *Stag* at Suez.

A new category of Wren Sick Berth Attendant replaced the VADs in October.

Mrs G. Pope was one of them:

I applied to join in August 1948 and was told that they were not recruiting any more VADs as the WRNS was to have its own Sick Berth Attendants. I was in the first class of Wren SBAs and was drafted with my class to Royal Naval Hospital Chatham... After six months' classroom training, we worked on the wards for a further six months. Following our final exams we were all drafted to various sick quarters on naval bases.

This tableau ended a review, starring Charlie Chester, at the Grand Opera House, Belfast in November 1950. Two minutes silence was observed during the tableau, the Last Post was sounded, and Mr Chester read Binyon's words. (Lent by Mrs Donaghy)

In December, Admiralty approval was given for a limited number of Direct Entries to commissioned rank. This scheme, open to women between $20\frac{1}{2}$ and 29, was intended to be of a temporary duration to meet a shortfall in officers.

Article in the April 1949 Issue of *The Wren*

The past year has seen some alterations and additions to the list of WRNS categories, notably the introduction of the Dental Assistant and Sick Berth Attendant. Another new category to be introduced is that of Radar Plot into which the Aircraft Direction category has been absorbed. WRNS Radio Mechanics have been renamed WRNS Radio, and Naval Airwomen have become WRNS Aircraft Mechanics; the latter will once more specialise as Air Mechanics (E) or (A) and will become 'Q.S.'. Leading Wren Aircraft Mechanics and above are now eligible to take the Air Engineer Officer's Writer's Course which, previously, has only been open to male Aircraft Mechanics.

On 1 February 1949 the Women's Royal Naval Service became permanent. The announcement was made by the Board of Admiralty in the following terms:

The W.R.N.S. is a permanent and integral part of the Naval Service and is regarded, in all respects other than its subjection to a separate disciplinary code, as part of the Royal Navy itself.

The outline rules were as follows:

The permanent Service has been planned to give a full career, whether officer or rating, to the woman who wants to make the Service her profession.

Ages of entry, 18 to 28. These limits allow ratings to complete the 22 years service necessary to qualify for pension before being obliged to retire at the age of 50.

The first engagement is for 4 years and will be followed by two more 4 year terms. Having completed 12 years in this way, ratings who are recommended may sign on for a final 10 years to complete time for pension.

Pensions and gratuities have been fixed at two-thirds of the men's rates and they are reckoned and awarded under conditions similar to those for men . . . The maximum rate of pension that can be earned by WRNS ratings is £3 a week, and the maximum gratuity £140.

Officers are given a commission and are deemed to have entered on an indefinite period of service for so long as they are required. Age limits for promotion are $20\frac{1}{2}$ to 29. At all stages promotion is by selection and none of it is on a purely time basis, but in each rank there have been established minimum and maximum ages and lengths of service. . . Officers must complete 20 years commissioned service to qualify for gratuity.

Both officers and ratings may claim discharge on marriage. . .

In April 1949 the Nursing Services were included in Women's Service Sport.

The Association of Wrens held their first post-war reunion in Westminster Hall on 22 October 1949.

CHAPTER 5
THE FIFTIES

The Fifties was a period of consolidation. Post-war austerities continued during the early years and the Service adjusted to new social attitudes. Many of the wartime categories disappeared, but new technology called for new skills. Memorial windows and the WRNS Book of Remembrance were dedicated. Dame Katharine Furse and Dame Vera Laughton Mathews, the first two Directors, died and two Wrens lost their lives in the ferry ss *Princess Victoria*. It was the decade of the New Look, a new Queen, and an expansion of WRNS employment into the new NATO Headquarters abroad.

The first WRNS Unit Officers' conference was held at HMTE *Dauntless*, on 11 May 1950.

MT Section HMTE Dauntless, *July 1949. (Lent by Miss E. H. Scott)*

Wrens on duty in the control tower, HMS Vulture. *(Lent by Mrs Broster)*

A stained glass window was unveiled by H.R.H. The Princess Elizabeth, on 29 October 1950, in St George's Church, Royal Naval Barracks, Chatham.

The window was given as a tribute to the WRNS in two World Wars by Chatham Wrens and ex-Wrens. It was designed by Mr Hugh Easton and portrays St Margaret of Antioch, the counterpart of St George.

A letter of appreciation for Dame Jocelyn Woollcombe DBE, who retired as Director WRNS 21 November 1950, was written by Vice Admiral Sir Michael Denny:

Dame Jocelyn assumed command of the WRNS at one of the most difficult and trying times that the personnel side of the Navy has gone through in the last century. It was a time when a rapid demobilisation was forced upon a structure whose post-war shape was far from clear. This was a particularly difficult period for the WRNS. It had no well tried peacetime organisation to fall back upon, and therefore had to start from scratch and construct, at very short notice, such an organisation to meet, not just the needs of 1946–47, but those of the next few decades should peace be maintained.

That this had clearly been achieved is the measure of the debt which the Navy owes to Dame Jocelyn. Under her strong lead and broad vision, the WRNS has been transformed from a war to peacetime footing without the loss of any of the characteristics for which the country holds it in such high esteem.   (Reprinted from *The Wren*)

In January 1951 a WRNS officer and five ratings were deployed to Norway to the Staff of the Commander-in-Chief, Allied Forces, Northern Europe.

A requiem Mass was held at Westminster Cathedral on 19 February 1951 for Dr Genevieve Rewcastle OBE, who was Medical Superintendent to the WRNS throughout the War.

In the October 1951 issue of *The Wren* an update on the Service said:

The WRNS Units at the Royal Naval Air Stations, Culham and Stretton closed last year. The WRNS were to have been withdrawn from HMS *Excellent*, but it was finally decided to retain a small Unit in this establishment. A Unit has been established at the Royal Naval Air Station in Malta. This is very welcome news to Wrens in the Air Command, as for the first time in the Permanent Service, Wrens serving in air categories now have a chance of going

overseas. One WRNS officer and five ratings are serving in Norway on the Staff of the Commander-in-Chief, Allied Forces, Northern Europe. There are now twenty-nine Units in the United Kingdom, two in Malta, one in Fayid, three WRNS officers in Germany, and the small party in Norway.

The two latest categories are those of Sick Berth Attendant and Dental Surgery Assistant. Both have proved very popular, and ratings serving in these categories are now doing valuable work in the Service. There have been no major additions or alterations in categories during the past year. The Advancement Regulations which were introduced in April 1949 and brought the WRNS into line as far as possible with the advancement of RN ratings, have become familiar, and the Service is gradually being looked upon as a worthwhile career.

The great event in 1950 was, of course, the new Pay Code introduced last September. This meant a substantial rise in pay, particularly for ratings, and although the initial stages caused a great deal of hard work for the Pay Writers, never, perhaps, was a task more welcome!

On 28 June 1952, eleven years after the loss of the ss *Aguila* and the 22 WRNS personnel who were *en route* to Gibraltar, the motor lifeboat *Aguila Wren* was launched at Aberystwyth. Dame Vera Laughton Mathews, wearing her wartime uniform of Director WRNS, paid tribute on behalf of the Service before introducing Mr Edward Benjamin, father of the late Chief Wren Cecily Benjamin. He, together with the late Canon Ogle, father of the Senior WRNS officer of the draft, were co-trustees of the *Aguila* Wrens Memorial Fund. In presenting the lifeboat to the RNLI, he said, 'that it was most fitting that these women who gave their lives should be commemorated in a boat whose sole purpose is to save life.'

Dame Katharine Furse GBE, RRC, the first Director WRNS (1917–1919), and President of the Association of Wrens died on 25 November 1952. A memorial service was held in St Martin-in-the-Fields on 16 December.

Wren H. J. F. Emerson sat for Gilbert Ledward RA who designed the Coronation five shilling piece and the Great Seal.

I had the honour to sit in for the Queen for the Coronation five shilling piece and the Great Seal of

Above: *WRNS Service Hockey Team, 1950.*
(Lent by Miss E. H. Scott)

Below: *The WRNS Contingent in the Coronation Procession on 5 June 1953.*

England. . . The original mould for the Great Seal was exhibited in the 60 years of portraits, and photographs of our present Queen, at the National Portrait Gallery.

Having played Queen Elizabeth I in the HMTE *Dauntless* production of 'Mirror to Elizabeth', First Officer Hooppell asked me if I would like to enter a competition being held by SSAFA for their Searchlight tattoo at the White City in 1952. It was open to all members of the WRAC, WRAF and WRNS. The scene was Tilbury before the Armada and we all had to learn the famous speech and then go to St John's Wood Barracks and ride side saddle on Winston (the horse the Queen always rode at Trooping the Colour). After being judged by (Sir) Richard Attenborough, Air Chief Marshal Sir Philip Joubert, Sir Michael Balcon and Googie Withers, I was declared the winner.

The spin off was reporting to Hammersmith Police Station for sittings on Winston for Mr Ledward, and I have his original sketches which he very kindly presented to me.

_____

Two Wrens were lost at sea when the ferry ss *Princess Victoria* sank between Larne and Stranraer on 31 January 1953.

A Wren Telegraphist trainee 1953. The site of the Signal School now forms part of the approach to the Tamar Road Bridge in Plymouth. (M.H.F.)

Three WRNS Officers (Cypher) accompanied the Royal Commonwealth Tour in November.

First Officer Joan Cole WRNS was loaned to Australia in November 1953. Her first appointment was to Flinders Naval Depot, about 47 miles from Melbourne, where she was Officer-in-Charge, WRANS, and responsible for recruit training. The Service only numbered about 300 and they were short of trained personnel. From the end of 1954, until she returned to the UK, she held the appointment of Director WRANS.

HMTE *Dauntless* was commissioned as HMS *Dauntless* on 11 December 1953. The actual celebration did not take place until 18 June 1954 when Dame Vera Laughton Mathews was invited to take the salute at a march past.

Dame Vera Laughton Mathews, wearing her wartime uniform inspects Divisions at the commissioning of HMS Dauntless. (Lent by Dame Nancy Robertson)

In January 1955 HMS *Wren* returned from eight and a half years abroad, mostly in the East Indies and Persian Gulf waters.

Before coming into Portsmouth a signal was made to the Chief Officer WRNS: 'On arriving in the Portsmouth Command after eight and a half years absence, your name-ship sends you and all the WRNS of the Portsmouth Command, greetings.' The Chief Officer signalled back, 'WRNS officers and ratings of the Portsmouth Command are delighted to welcome HMS *Wren* back to Portsmouth.'

Above: *H.R.H. The Duchess of Kent inspects Divisions during a visit to* HMS Dauntless *in 1954. (Lent by Dame Nancy Robertson)*

Right: *Dame Vera Laughton Mathews was present when* HMS Wren *returned to Portsmouth. She had launched the ship in 1942. (Lent by Mrs Broster)*

On 22 October 1955, Director WRNS accepted a portrait of Dame Katharine Furse by Miss Marcelle Morley (ex Second Officer WRNS). The painting hangs in WRNS quarters at Furse House, London.

In 1956 the minimum age for entry was reduced from 18 to $17\frac{1}{2}$. This was the second time in the WRNS history that the minimum age was reduced to under 18. Time served prior to 18 did not count towards pension and gratuity entitlement, and parental consent was necessary for under-18 applicants.

WRNS personnel increasingly were being employed in NATO headquarters. Shortly after joining

Above: *One of the last Officer's Training Courses to wear black tricorns.* (Lent by Mrs Broster)

Below: *The first Officer's Training Course to wear permanent white cap covers.* (Lent by Mrs Broster)

No fuel during the Suez Crisis – the captains of HMS Excellent *and* HMS Vernon *opt for alternative transport. (Lent by Miss E. H. Scott)*

the staff of Commander-in-Chief Allied Forces Mediterranean, Malta, Third Officer M. V. Jackson accompanied an Exercise Planning Team to Naples, Athens, Ankara, Bizerta and Gibraltar. She was secretary to the team and Leading Wren C. J. Watt was the stenographer.

On 1 November 1957 the First Lord of the Admiralty addressed the WRNS reunion.

Today we have only 3000 Wrens; about the same number as in this hall tonight. But they form a vital part in the structure of the Royal Navy; and, indeed, in a good many categories they fulfil their role better than men. This is increasingly necessary today when, with the reduction of our naval strength, it is essential to send every possible Royal Naval officer and rating to sea if we are to maintain a fleet of respectable size. Wrens, therefore, have a vital role to play in filling jobs ashore so that more ships can be manned at sea. . . I can assure you that the Board of Admiralty has given no thought at all to the possibility of dispensing with the services of the WRNS and, so far as I am concerned, I do not intend to give the matter one moment's consideration.

The RNVR became part of the Unified Reserve on 1 November 1958 and dropped the word 'Volunteer' from its title. Similarly the WRNVR dropped 'Volunteer'.

From 21 November 1958 Night Shore Leave for Leading Wrens and Wrens over 21 was extended to midnight.

The Book of Remembrance containing the names of all who died while actually serving in the WRNS from 1917–1919, and from 1939 onwards, was dedicated at the Royal Naval College Chapel at Greenwich on 14 March 1959.

The ceremony of dedication was very simple and

The WRNS unit, HMS Excellent, *1957.*
(Lent by Miss E. H. Scott)

The end of the Secretarial Course at HMS Ceres,
*1957. (*M.H.F.*)*

dignified. The beautiful dark blue book was carried up the aisle on a cushion by Chief Wren Glory England with two WRNS supporters, and received at the altar steps by the Director WRNS, who asked the Chaplain of the Fleet to dedicate it.

Venerable Sir, we pray you to receive and dedicate to the Glory of God this Book of Remembrance which contains the names of all ranks of the Women's Royal Naval Service who have died while serving their country in war and peace.

He, taking the Book, replied, 'I am willing so to do'. In his Address, the Chaplain of the Fleet said.

This is the day of which and in which to be proud. A day to remember all those who have died whilst actively serving in the WRNS.

The RNAS Eglinton WRNS Rifle Team at the Rifle Championships, Ballykinlar, 1956. (M.H.F.)

WRNS Sportswomen of the Year, 1956. (Lent by Miss E. H. Scott)

There are just over three hundred names in this Book of Remembrance. That is not a very large number when you consider the total number of Wrens there have been... One remembers particularly those nineteen lost at sea on August 19, 1941.

Some of those we remember today may have been known to you, possibly personal friends, and others unknown, yet, while we should think of them individually as persons and commend their souls to the loving mercy of our Heavenly Father, we shall also think of them as typifying the whole spirit of all those who served in the Women's Royal Naval Service.

A very young Service in the matter of years but a Service which in a short period of time has acquired a completeness, a maturity, a spirit and a tradition which has made it a recognised and accepted part of the Service, as a whole. It would, I think, be difficult now to visualise the Navy without the WRNS...

The First World War gave a glimpse of the possibility but the Second enabled women to prove to themselves as well as to the world not only what they could do, but in fact, how very little there was they could not do.

My own recollection of the WRNS right from the beginning of the War was a tremendous spirit of enthusiasm, a pride in and devotion to the Navy and a single-minded sense of duty.

That is something of which you can rightly be proud and because you are proud of that tradition, and of those who helped to build and uphold it, you have felt an instinctive wish to enshrine that

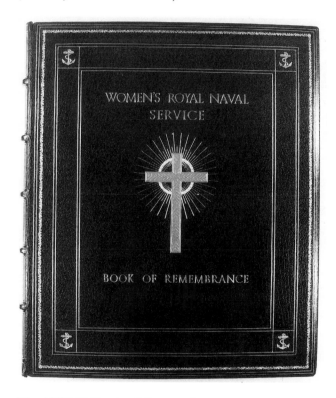

The WRNS Book of Remembrance.

tradition and perpetuate their memory in this beautiful Book of Remembrance which has just been dedicated, and finds an honoured place in this historic Chapel.   (Reprinted from *The Wren*)

Dame Vera Laughton Mathews, DBE, Director WRNS (1939–46), and President of the Association of Wrens, died on 25 September 1959.

CHAPTER 6

THE SIXTIES

The Sixties opened with the 21st anniversary of the re-formed Service, to be followed in 1967 by the Golden Jubilee of the first formation of the WRNS. New accommodation was built, terms of commissions and engagements were amended, and WRNS personnel deployed to Mauritius.

Her Royal Highness Princess Marina, who had been Chief Commandant throughout the war years

WRNS Officers in Malta celebrate the WRNS 21st Birthday. (M.H.F.)

and during the subsequent peace, died towards the end of the decade.

In 1960 the Wren Sick Berth Attendant category which had superseded the VAD in 1948, was in turn superseded when the Queen Alexandra's Royal Naval Nursing Service introduced ratings into their corps. One of the last Wren SBAs to be trained was Avril O'Donnell who wrote in *The Wren.*

As the train rumbled into Gillingham station doubts began to rush into our minds. We were nine

brand new Wrens starting our Part II training as Wren Sick Berth Attendants at the Royal Naval Hospital, Chatham. The last of the Wren SBAs in the Royal Navy, as the category was being discontinued.

It was the 21st anniversary of the re-formation of the WRNS on 11 April 1960, and throughout the year various events took place to celebrate the WRNS coming of age.

On 5 November 1960 the Association of Wrens held a 21st birthday reunion, in the presence of Her Majesty Queen Elizabeth the Queen Mother. During the reunion the first public performance was given of J. Vivian Dunn's 'March of the WRNS' which had been approved as the official march on 11 October. In her address Her Majesty said:

Your reunion this evening marks an important anniversary for the Women's Royal Naval Service in its present form. For you it is a twenty-first birthday and I offer you my warmest congratulations and am delighted to be with you to celebrate the occasion and to wish you many happy returns of the day. . .

When in 1939 I became Commandant-in-Chief, it was my earliest pleasure to invite Her Royal Highness The Duchess of Kent to make the WRNS her special interest . . . I have seen the Cooks at work in the huge galleys of the Naval Barracks; I have watched the Plotters performing miracles of skill in their close, underground quarters; I have tried to fathom the mysteries of the Radio Mechanics and Torpedo Wrens. My most vivid memory is of the day in Devonport in 1942 when I travelled from North Dockyard to the Naval Barracks pier in a Vosper speed boat, because that was the first occasion on which a Royal Barge was ever manned by a Wrens' Boat Crew, and an occasion I certainly will not forget.

This evening you will be meeting old friends and recalling both the dangers and discomforts which you shared and the amusing incidents which lightened the dark days of the war . . . All of you must have gained much from your days in the Service. . .

During two World Wars and in times of peace, often fraught with anxiety, your standard of loyalty and devotion has never varied. In the years ahead I am confident that this splendid tradition will be upheld and I offer you all my warmest good wishes for the future.   (Reprinted from *The Wren*)

In the Lord Mayor's Show on 12 November 1960, the WRNS float commemorated the 21st birthday of the Service.

WRNS personnel, in the course of their duties were, by this time, being employed in ever widening ranges of tasks. Leading Wren G. Johnson, as a Writer (Shorthand) on the Staff of Earl Mountbatten of Burma, accompanied him during a world tour in October 1961. The trip took in El Adam in Tripolitania, Aden, the Maldive Islands, Malaya, Singapore, Bangkok, Darwin, Canberra, Wellington, Honolulu, San Francisco, Ottawa and home. In her report for *The Wren* she commented:

Most of our dictation and typing was being done at a height of 35,000 feet and at a speed of 500 miles per hour. They didn't tell us that this might happen during our training days in HMS *Pembroke*.

The Dame Vera Laughton Mathews' Award was instituted from the Memorial Fund in June 1961. The Award Fund was in the custody of the WRNS Benevolent Trust and the award would be made annually. The purpose was to enable the daughter of an ex-WRNS officer or rating to take full advantage of higher education, or special training, likely to develop her character and potential capabilities. The Award is now managed by the Association of Wrens.

H.R.H. The Duchess of Kent visiting RNAS Lossiemouth, 1961. (M.H.F.)

Ante-dated seniority was introduced on 8 October 1962, for Meteorological Officers with Short Service Career Commissions, and Direct Entry Officers for Permanent Commissions.

On 1 February 1963 an obligatory subscription of

1/- was introduced, payable to the WRNS Central Sports Fund, to cater for wider participation.

In October 1963 the age for entry to the Quarters Assistant category was reduced to 19.

Until this time the minimum age on entry was 21 as it was for the new entrants accepted for the Regulating branch. As soon as these Wrens had completed their professional training they were advanced to Acting Leading Wren.

Leading Wren Diana Mann, Writer (Shorthand) was the first serving member of the WRNS to achieve the Gold Standard in the Duke of Edinburgh's Award Scheme.

In 1964 First Officer O. D. Middleton was the first WRNS officer to specialize in Work Study.

At 2359 on 31 March 1964 the last Admiralty General Message was made. It read:

> The Lord High Admiral's Flag was lowered with due ceremony this evening from the Admiralty in the presence of the Lords' Commissioners. This same flag will tomorrow be presented to Her Majesty, who has graciously directed that responsibility for its future safe keeping shall rest in the hands of the Admiralty Board. The Commissioners for executing the Office of Lord High Admiral wish to assure all in the Naval Service that the Admiralty Board will in the future hold their interests and well being equally in mind.

The Board Room remains as it was and continues to be used for officers' promotion boards, including the WRNS.

On 4 May 1964, Seven-Year Short Service Commissions were introduced for WRNS officers.

On 29 May the age on entry for WRNS ratings was reduced to 17.

On 1 October 1964 the Six-Year Engagement and Bonus Scheme was introduced for WRNS ratings.

On 30 January 1965 the Director WRNS and the Matron in Chief QARNNS were amongst the congregation in St Paul's Cathedral for the funeral service of Sir Winston Churchill. The WRNS and QARNNS, represented by a contingent of two officers and fifty ratings and naval nurses 'Kept the Ground' in the vicinity of the Cathedral.

The RNLI lifeboat *Aguila Wren* was transferred from Aberystwyth to Redcar, Yorkshire lifeboat station in February 1965.

On the occasion of the 25th anniversary of the appointment of Her Royal Highness Princess Marina, Duchess of Kent, as Chief Commandant of the WRNS, the following message was sent on 11 March 1965:

> On the eve of the twenty-fifth anniversary of Your Royal Highness' appointment as Chief Commandant of the Women's Royal Naval Service, the Admiralty Board and all officers and ratings of the Service convey their warm congratulations, together with an expression of their deep appreciation of Your Royal Highness' continuing interest in all that affects the well being of past and present members of the Service.

The following reply was received:

> I write to thank you for your kind letter concerning the 25th Anniversary of my appointment as Chief Commandant of the Women's Royal Naval Service.
>
> Would you please convey my gratitude to all officers and ratings of the Service who joined in sending me this message of appreciation, which has greatly touched me.
>
> It has been a source of much pride to me that I have been associated for so long with a Service which continues to maintain such a high standard and to command the respect of everyone.
>
> Marina

In Spring 1965 a new branch was opened to graduates with a degree in Electrical Engineering, and equivalents, to specialize as Air Electrical Officers.

Early in 1966 training as Air-Traffic Controllers was opened for WRNS officers.

Similar duties had been carried out by WRNS officers during the Second World War, but this requirement had lapsed in post-war years.

RPO Wren D. A. Watkinson was the first winner of the NATO Sports Trophy – for athletic achievement in 1965.

In May the first detachment of WRNS personnel arrived at HMS *Mauritius*. Mrs M. G. Robinson was one of the original draft.

> One Chief Wren Radio supervisor, one Acting Leading Wren Radio Operator, four Able Wren Radio Operators (M) and four Able Wren Radio Operators sat excitedly in the departure lounge at Cromwell Road bus station. . . We were the first

Wrens to be drafted to HMS *Mauritius* and we were all very aware of just how lucky we had been in being picked for the initial six months' draft.

I spent one very happy year on that paradise island and often look back to the memories of sea, sand and sunshine, not to mention the cyclone which occurred; we were amazed to see ropes leading from mess deck to place of work; needless to say we soon discovered why – the wind was strong enough to blow you right over. . .

Two Mauritian ladies were employed to clean our living quarters and to serve in the dining room . . . When Chief Wren announced that the Director was coming to visit, Andre and Simone were very impressed – indeed we had a very hard job convincing Simone that Director WRNS was *not* Her Majesty the Queen.

On 8 June the WRNS was represented at the Dedication, in Cardiff, of the Welsh National Book of Remembrance in the presence of Her Majesty Queen Elizabeth The Queen Mother.

The first post-war draft of WRNS personnel left for Singapore on 3 November.

Her Royal Highness Princess Marina presided at the 25th Annual General Meeting of the WRNS Benevolent Trust on 29 April 1967. In her speech Princess Marina said.

It is strange to reflect that as long ago as 1943 I presided at the first Annual General Meeting of the Trust, when the bye-laws had been in operation for only a few months and were still the subject of amendment in the light of experience. What a lot has happened since those days. . . In the intervening years the work of the Trust has steadily grown and expanded, continuously moulding its policy to meet the demands of changing times. . .

During the period of its existence, the Trust has helped 6,135 members. Each year has brought an increasing number of requests and it is now a matter of urgency to seek out those needy members who are unaware of the help that is available to them. The wartime Wrens are not getting any younger – the pattern is changing and higher maintenance grants will have to be given in the future. It is therefore imperative that the funds be kept buoyant to meet demands, so that we can look forward to the present number of beneficiaries being more than doubled in the next 25 years. The machinery for this exists and as the years go by I feel

sure that the aims of the Trust can be attained. Everyone interested in the Trust can help even in a small way.

1967 also saw the Golden Jubilee celebrations of the first formation of the Women's Armed Forces.

There was a service in Westminster Abbey attended by Her Majesty Queen Elizabeth The Queen Mother and the three Chief Commandants; a garden party at Burton Court attended by Her Majesty The Queen, His Royal Highness The Duke of Edinburgh, Her Majesty The Queen Mother and the three Chief Commandants; and a dinner was held in the Painted Hall at the Royal Naval College, Greenwich for officers of all three Services.

The Golden Jubilee WRNS reunion was held in the Festival Hall on 27 July 1967, in the presence of Her Royal Highness Princess Marina.

A *Tableaux Pageant*, 'Fifty Years On', written by Mrs Esta Foster (née Eldod), an ex-WRNS officer, was performed at the reunion. In her speech Princess Marina said:

Fifty years ago! What a long time it seems – and indeed it is. Most of you in this great hall tonight, I rather think, were not even born – let alone in your cradles – when, in 1917, the WRNS were formed. Quite a lot of us on this platform, however, were well on our way by that time! I am so glad to hear that there are some Wrens with us tonight who joined in those days; we are very proud to have you with us and to think that it was our privilege, in the Second War, to follow in your footsteps.

How well I remember the day in 1939 when my brother-in-law, the late King, appointed me your Commandant. Naturally – though apprehensive – my first action was to ask that very great and remarkable person, Dame Vera Laughton Mathews to come and see me at Buckingham Palace where I was staying at the time. . . it was with some trepidation that I, as a mere fledgling, looked forward to this interview. While in theory it was I who was to interview dear Dame Vera, in the event there was no doubt whatever that it was she who interviewed me!

I think it was soon afterwards, and at Admiralty House, that I committed the appalling solecism of wearing white gloves! And I rather fear that this was not the last time that, quite unwittingly, I infringed the Regulations for Officers as laid down in Admiralty Fleet Orders. Cap covers – or not cap

H.R.H. The Duchess of Kent and her 'white gloves'.

covers – were also sometimes my undoing. And I must confess that it took me a little time in those early days to get used to calling a kitchen 'a galley' and a bedroom 'a cabin' and I am afraid I slipped up occasionally. . .

I am proud to have been connected with this great Service as its Commandant for very nearly 30 years. It has been a privilege, indeed a great honour, and a source of much happiness to me as well. And so to all of you Wrens, past and present, I would like to express my thanks, together with my affectionate good wishes.

Early in 1968 Cadet Entry was introduced.

During 1968 three Chief Wrens were working in the Family Welfare Office of HMS *Terror*, Singapore whose combined service totalled 81 years. They were Chief Wren Violent Perrin who, at the age of 62 was the oldest serving Wren and retired in May 1968; Chief Wren Hilda Earl, aged 55, the longest serving Wren who first put on uniform in 1939; and Chief Wren Muriel Gould aged 51.

The minimum age limit for promotion to Probationary Third Officer was reduced to 20 on 29 July.

At the Festival of Remembrance in November 1967, WRNS personnel appeared in a tableau which included the changing uniform.

Her Royal Highness Princess Marina died on 27 August 1968.

Dame Jocelyn Woollcombe said in a tribute in *The Wren.*

On the outbreak of war she was appointed Commandant of the WRNS, a title which was later to be changed to Chief Commandant. From the beginning she was determined to learn all about the Service; not least did she have to learn, as we did, to wear uniform correctly! She reminded us last year at our Reunion how difficult this was for her, and I remember well the expressions of shock and amusement mingled in her face as she stepped from her car and realised that she and the reception party were at variance about the wearing of white cap covers, and of the carrying of non-Service handbags. But, wherever she went during those war years – to Air Stations, to the Home Ports or to isolated small units – the sight of her was a tonic and her genuine unfailing interest in what we were doing and how we lived was an inspiration. . .

During the War, in common with so many other women, she suffered the tragic loss of her husband; but we were not allowed to suffer. Quietly she assumed many of the Duke of Kent's social obligations, untiringly she came amongst us, and as our duties and categories expanded, she learnt more and more of what we did and where we served. I wonder how many of us know that she also spent much of her time as an anonymous VAD in University College Hospital.

When the WRNS Benevolent Trust was launched she graciously consented to become its President, and when the Association of Wrens of World War I opened its doors to the Wrens and ex-Wrens of World War II, she became our Patron. . .

Her last message to the Association, written in her own hand on her Christmas card, sums up her feeling for us. She wrote, 'With my affectionate good wishes to you – Marina'.

A Memorial Service was held for Princess Marina in Westminster Abbey on 25 October.

In the subsequent issue of *The Wren* appeared the following.

Every seat in the Abbey was taken and it was good to see so many Wrens, both old and young, in that congregation of the famous – representatives from the Association of Wrens and some of the Branches, and from the Benevolent Trust; the ex-Directors; and a heartening number of smartly uniformed serving Wrens, led by the Director who sat with the Second Sea Lord. Several WRNS officers acted as ushers.

WRNS personnel attended the Investiture of H.R.H. The Prince of Wales at Caernarvon on 1 July 1969.

It was announced at the AGM, on 25 October, that the Association of Wrens had received charitable status.

A WRNS officer joined the Staff of Commodore Hong Kong in October 1969, and the first Wren Air Hostess was employed by the Fleet Air Arm.

Second Officer Carol Sturgin was one of the first WRNS officers to be exchanged with the US Navy. She was appointed to the Training School at Bainbridge, Maryland.

CHAPTER 7
THE SEVENTIES

This decade was to see major alterations with regard to the organization and administration of the WRNS and, in some respects, the Armed Forces as a whole. Her Royal Highness The Princess Anne became Chief Commandant, the Service came under the Naval Discipline Act and WRNS officers received the Queen's Commission. Study groups reported on the 'Way Ahead' and the future of the Service, and organization and administration was closely integrated with the Royal Navy. The Diamond Jubilee of the first formation of the Women's Services was celebrated, and exhibitions were held at the Imperial War Museum and the National Maritime Museum.

On 20 January 1970 the Association of Wrens celebrated its Golden Jubilee.

Admiral of the Fleet the Earl Mountbatten of Burma wrote in his Foreword to the Jubilee issue of *The Wren*.

The Women's Royal Naval Service originated in the First World War and reached the strength of 438 officers and 5,954 ratings. . .

During the Second World War the Women's Royal Naval Service expanded to 5,061 officers and 69,574 ratings under their Director who was granted the equivalent in rank to a Rear Admiral. Officers and ratings held key jobs both at home and abroad. . .

Although in peace time the strength of the Navy is very much less than in war, proportionately the officers and ratings of the Women's Royal Naval Service have maintained a very important role.

The WRNS continues to play a prominent part in the contribution which the Royal Navy makes towards the defence of the United Kingdom, the British Commonwealth and the North Atlantic Treaty Organisation.

Past and present members of the WRNS can be justly proud of their great contribution towards victory in two World Wars. Their serving successors of today are carrying out their role splendidly and are carrying on the great traditions established by their predecessors.

To all Wrens, officers and ratings, past and present, I send my very best wishes on this the 50th anniversary of the formation of the Association of Wrens.

In January a WRNS officer was appointed to the Staff of the Director of Public Relations (Navy).

The Military Salary was introduced in April.

Night Shore Leave, which had been extended to midnight in 1958, was amended to All Night Leave, the same as the men.

Her Majesty Queen Elizabeth The Queen Mother was present at the Golden Jubilee reunion of the Association of Wrens on 30 October.

The following message was received from Her Majesty The Queen.

Please convey my warm thanks to all Members of the Association of Wrens for their kind and loyal message.

I send my sincere congratulations to the Association on the celebration of your Golden Jubilee with my best wishes for the years to come.

Elizabeth R.

Dame Jocelyn Woollcombe, President of the Association, was unable to attend, because of illness, but her message of welcome was read by Dame Mary Cheshire in which she said:

Ten years ago we celebrated in this hall the 21st birthday of the Women's Royal Naval Service of 1939, and we had the great honour and happiness of welcoming you, Ma'am, to that reunion.

Tonight we come with, if possible, even warmer hearts to welcome Your Majesty to another anniversary, the Golden Jubilee of our Association. . .

Since our years of service we have followed many different paths. Probably most of us have married, and not a few have daughters in the WRNS. We come tonight from many different places to renew old friendships, to tell again old stories of the past we shared, and in so doing renew our youth. What matter, Ma'am, if the waistline is a little larger, the fingers less skilful, the pace of life a little slower! But it is wonderful how the years fall away at a reunion. . .

The pride, the happiness and the gratitude that we felt then at the thought of welcoming you, we feel again tonight. Perhaps we have not risen at dawn to scrub the galley and tidy our quarters – though I think that some have indeed risen before dawn to be here. But, Ma'am, the enthusiasm is the same and we welcome your gracious presence amongst us with all our hearts.

In her reply, The Queen Mother said:

Once again you have come together to celebrate an anniversary – The Golden Jubilee of the Association itself. This is indeed a matter of pride, for Associations do not live for half a century unless they are really worthwhile.

Fifty years ago, in 1920, the Wrens of World War One were all back in civilian life, but their great leader, Dame Katharine Furse, believed that they would want opportunities to meet and talk of shared experiences. . . One of the people who served under Dame Katharine in the first War, who was later to lead the newly-born WRNS in World War Two, was Dame Vera Laughton Mathews. Over the years I have many vivid and cherished memories of visits to Wrens in all sorts and conditions of places. In barracks Wrens were learning all the facets of a Writer's job so that men might be spared to go to sea. I also saw Radio Mechanics, Shipwrights, Signallers and Plotters at their work, and I always came away filled with thankfulness and pride at the loyalty and devotion to duty shown by these splendid people. . .

An anniversary is a time to look back, but it is also a time to express wishes for the future. I wish for you all that your Association may increase in numbers and grow in value, forming a meeting place where old and young may come together and so carry into civilian life the spirit and traditions of the Service they have shared.

On 1 January 1971 Naval Pay was computerized.
In May 1971 a Leading Wren Writer (S) became PA to the Defence Adviser, British High Commission Wellington, New Zealand.

In July 1971, the first WRNS personnel were drafted to the NATO Headquarters in Lisbon, Portugal.

On 30 September Chief Wren Beatrice Willis was the first selection for the newly introduced rank of Fleet Chief.

On 1 May 1972 the Nine-Year Notice Engagement was introduced for WRNS ratings.

A new category of Photographer was introduced in June 1972.

In April 1973 the first Stores Assistants qualified as PO Wren Stores Accountants.

By this time the various strands of the Stores category had, yet again, been reorganized. Other categories were also being reviewed and brought up to date as technology enhanced efficiency and removed the need for some of the manual tasks.

The lifeboat *Aguila Wren* was handed over to the Sea Cadet Corps at Keadby Lock, near Scunthorpe on 20 May 1973. The lifeboat, presented to the RNLI from the Memorial Fund raised by relatives of the Wrens lost at sea in the ss *Aguila* in 1941 had now completed her useful life in that capacity. She served in Aberystwyth between 1951 and 1965 when she was transferred to Redcar. During her service as a lifeboat she was launched 52 times and saved a total of 36 lives. Normally when lifeboats reach the end of their active service they are sold without their livery and name. In this case, however, the *Aguila Wren* has retained her RNLI colours as well as her name in order to preserve the memorial to the Wrens who lost their lives. She is now known as the TS *Aguila Wren*.

In July 1973 a working party reported on conditions of service for WRNS Officers.

On 26 January 1974 the Central Committee of the WRNS Benevolent Trust replaced Command Committees.

This was a natural rationalization of the Trust as the Naval Commands altered in size and shape. The Commands, subsequently, provided representatives to serve on the Central Committee.

In January 1974 WRNS officers were again appointed to the Royal Naval Staff and Lieutenants' Courses at Royal Naval College, Greenwich. Although WRNS officers had attended these courses during the war, they had lapsed for many years. By the Seventies, the professional training of WRNS officers, together with the types of appointments they were filling, called for selected officers to be staff

trained. Today, most WRNS officers receive progressive staff training during their careers which may be through the Lieutenants', Naval Staff, or National Defence College courses; or a combination of all three.

On 1 February 1974 the Director WRNS received the following letter from the Secretary of State for Defence.

On the occasion of the 25th anniversary of the reformation and incorporation of the Women's Services as regular forces of the Crown, I offer my congratulations and best wishes to all members of the Women's Royal Naval Service.

Their work is admired and much appreciated by all concerned.

On 1 July Her Majesty The Queen approved the appointment of Her Royal Highness The Princess Anne as Chief Commandant WRNS.

The Director WRNS called on H.R.H. The Princess Anne on 11 July, and the Chief Commandant lunched informally with the Senior WRNS Officer on 29 July. The Princess Anne became Patron of the Association of Wrens on 9 August. Her first official engagement as Chief Commandant was when she took the salute at a passing out parade at HMS *Dauntless* on 22 October 1974.

In November the report on the 'Future Role, Structure and Organisation of the WRNS' was completed.

On 27 January 1974 HMS *Amazon* was adopted as the WRNS ship, replacing HMS *Wren* which had completed her naval life.

From 4 March 1975 WRNS ratings, who had served for one year, were allowed to live ashore under the same rules as naval ratings. H.R.H. The Princess Anne attended a WRNS officers' reception at St James' Palace on 17 March 1975.

The WRNS team won the Best Ladies Team Trophy, and Leading Wren Precious won the Best Lady Driver award, at the Driver of the Year competition at RAF *Uxbridge* on 1 October 1975.

A Leading Wren Writer was drafted to the Naval Attaché's Office, Peking in November 1975.

On 1 January 1976 the Nine-Year Notice Engagement replaced the Career Engagement.

After basic and professional training, and a speci-

Her Royal Highness The Princess Anne became Chief Commandant WRNS, in July 1974.

fied period in the Service, ratings could now give eighteen months' notice of their intention to leave the Service before the end of their contract. Rules regarding discharge on marriage were not affected.

Her Royal Highness the Princess Anne took the salute at the WRNS OTC passing out parade at Royal Naval College, Greenwich on 19 March. This was the last course to be trained at Greenwich prior to the transfer to Britannia Royal Naval College, Dartmouth.

On 8 April BR 1077 'WRNS Regulations' became obsolete. The decision that the WRNS should come under the Naval Discipline Act removed the need for a separate set of regulations, and work had been in hand for many months to incorporate the necessary amendments to Queen's Regulations for the Royal Navy.

In September 1976 the first course of WRNS officer candidates arrived at Britannia Royal Naval College, Dartmouth. WRNS officers' training had been carried out at the Royal Naval College Greenwich since 30 October 1939, except for a period during World War Two when it was necessary to evacuate because of bomb damage. Over the years the course had developed from a basic two weeks to three months and was aimed at preparing officers for duties and responsibilities in connection with the WRNS own organisation and administration. Upon leaving Greenwich a proportion of officers would receive professional training in specific branches.

By the Seventies the training had been expanded to take account of the wider Naval responsibilities WRNS officers were assuming, but there was a limit to what could be achieved with the WRNS officers isolated from the Naval Officers Academy and the decision was taken to transfer to Dartmouth.

WRNS ratings Overseas Boards were abolished in November. Until this time WRNS ratings considered for overseas duty had appeared before a selection board. In keeping with their increasing professional expertise, and the forthcoming changes in the organisation and administration of the Service, steps were being taken to rationalize the management of the Royal Navy. Drafting, including overseas, was one of the early amalgamations. Recruiting and entry procedures, advancement rules, officers' promotion and appointing were to follow in due course as the Royal Navy progressed towards the maximum effective use of its personnel.

The first WRNS OTC at Britannia Royal Naval College, Dartmouth.

Portrait of H.R.H. The Princess Anne now hung in Dauntless *Squadron at* HMS *Raleigh.*

On 2 December Her Majesty Queen Elizabeth The Queen Mother unveiled a portrait of Her Royal Highness The Princess Anne at HMS *Dauntless*.

In March 1977 the Combined Women's Services Diamond Jubilee was commemorated by a service in Westminster Abbey. The service was attended by Her Majesty The Queen Mother and the three Royal Commandants.

The Second Officer Pauline Doyle Trophy for 'The greatest prowess in Parade Training without losing femininity' was presented to the WRNS OTC at Dartmouth in March. The first 'Young Sportswoman of the Year Trophy' was presented to Wren S. J. Sim.

In June educational standards for promotion to officer were altered to equate with the Royal Naval requirements.

On 28 June WRNS personnel were present at the

Opposite: HMS **Amazon** *replaced* HMS **Wren** *as the WRNS ship.*

Silver Jubilee Review of the Fleet at Spithead.

On 1 July 1977 the WRNS came under the Naval Discipline Act. The WRNR celebrated its 25th anniversary on 20 July.

On 7 September there was a service to commemorate the Diamond Jubilee of the WRNS.

A reception for officers and ex-officers of the Women's Services was held in St James's Palace on 27 October.

A WRNS Commemorative Stamp Cover was issued in November. An Exhibition, 'W.R.N.S. 1917–1977' was opened by Lord Mountbatten at the National Maritime Museum on 2 November.

A new WRNS officer career structure was introduced on 1 January 1978.

On 3 January a new category of WRNS PT & R was introduced.

Chief Officer Hilary Jeayes, OBE, Commanding Officer of HMS *Dauntless*, named locomotive 50048 *Dauntless* at Reading railway station on 16 March.

The Pauline Doyle Trophy.

Miss Ursula Stuart Mason (left) and Director WRNS at the entrance to the WRNS Exhibition.

WRNS personnel volunteered and deployed with 42 Commando Royal Marines, for duty in Northern Ireland between May and November. Out of two hundred or so applications, seven Wrens were chosen. This was the start of a practice which continues today. Whenever a Royal Marine Commando deploys to Northern Ireland a small contingent of Wrens, with a Third Officer WRNS in charge, provides support.

In 1978 the Family Services category came into being which incorporated the WRNS Welfare category.

An example of the increasing use of WRNS personnel to maximum effectiveness, wherever they were needed, is the following extract from an article written by Leading Wren Wilson – a Radar Plotter – for the *WRNS Newsletter*.

I had come to America as part of an analysis team from Northwood, sent to work in conjunction with the American Navy analysing and writing reports

Opposite above: Wrens 'splice the mainbrace' to celebrate the Silver Jubilee.

Opposite: H.R.H. The Prince of Wales presenting 'wings' to the first Wren Cabin Attendants at Yeovilton 20 September 1977. Left to right: Leading Wren Karen Nelson, Wren Isabel Gower, Wren Helen Watson.

Above: *H.R.H. The Princess Anne visiting* HMS Cochrane, *20 October 1976.*

Below: *Wrens with 42 Commando, Royal Marines in Northern Ireland, 1978.*

Happiness is serving with 42 Commando.

on a recent NATO exercise. . . As a Radar Wren employed in the Operational Analysis Group at Northwood, it was my job to draw the necessary diagrams and plots for the final report. I also learned the new skill of digitising, something I thoroughly enjoyed and which made a change from the drawing board. The job was most worthwhile and particularly rewarding when I saw my plots and graphs in print.

Wrens serving with 41 Commando Royal Marines in Cyprus, between May and November 1979, received the United Nations medal for peacekeeping duties.

By 1979 all appointing for WRNS officers had been transferred from Director WRNS to the Naval Secretary's Department. A Chief Officer WRNS had been added to his complement. In June the Naval Secretary also took responsibility for the appointment of Superintendent WRNS.

A WRNS and Association of Wrens window, designed by Lawrence Lee, was unveiled in the Cathedral of the Holy Spirit, Guildford, in the presence of Lord Mountbatten, on 7 July.

On 30 July WRNS Air Mechanic (Airframes and Engines) commenced training as Aircraft Mechanicians.

Her Royal Highness Princess Anne presented a 17-foot cabin cruiser to HMS *Dauntless*.

The cruiser was to be used on the local canal system until 1981 when HMS *Dauntless* transferred to HMS *Raleigh*. As the cruiser was not suitable for coastal waters it was sold and the money used to buy an antique table and pictures for the New Entry accommodation.

The Wrens in Naples visit their American counterparts on board USS *Vulcan.*

Opposite above: Just one facet of being an Air Mechanic.

Opposite: Wren Radar 'teaching' Weapons Officers at the School of Maritime Operations.

CHAPTER 8

THE EIGHTIES

This decade has seen the implementation of decisions leading to the closer integration of the WRNS with the Royal Navy. Cuts in manpower, and the need to make the maximum effective use of all naval personnel ashore to support the Fleet, have broadened the scope of employment of WRNS personnel. As they receive identical professional training in the branches which provide shore support, Wrens are interchangeable with their male counterparts. This was particularly evident during the Falklands crisis when WRNS personnel were moved, at short notice, to take over the operations cells in order to release men for sea duty.

The process of integration continues and it is normal today for posts to be nominated male/female, for WRNS personnel to be attached to ships in refit and in the shore support team for seagoing squadrons, to serve at sea for short periods with helicopter units, man Operations Rooms, serve with the Royal Marines in Northern Ireland and, in fact, respond to any task ashore.

Wrens work in all shore establishments in the UK and are also in evidence overseas, particularly in the NATO environment in Belgium, Italy, Norway, Denmark and Portugal. They are on exchange in the United States and serve in Hong Kong, Diego Garcia, Gibraltar, Northern Ireland and the Falkland Islands. In addition there are individual exchanges, periodically, with the Army and Royal Air Force.

In January 1980 a Chief Officer WRNS was appointed to the Directing Staff of the Royal Naval Staff College.

In August 1980 management of overseas drafts was transferred from HMS *Dauntless* to the integrated Royal Naval organization.

In October 1980 WRNS officers joined the duty list as Officer of the Day.

The first WRNS Air Engineer Officer started training at the Royal Naval Engineering College, Manadon in January 1981. She was a Graduate Entry Officer and subsequently served at HMS *Heron*, Royal Naval Air Station, Yeovilton.

A WRNS Staff Officer was appointed to the Sea Cadet Corps in March 1981.

The Girls Nautical Training Corps was disbanded and girls were admitted to the Sea Cadet Corps – to be called Nautical Training Contingent. As the Sea Cadet Corps had Ministry of Defence support, and appointed a number of naval officers for staff duties, it was appropriate that a WRNS officer should be included in the management structure.

In April 1981 a working party reported to the Admiralty Board on 'The WRNS in the 1980s'.

HMS *Dauntless* was transferred to HMS *Raleigh*, on 15 August 1981, where it would be known as *Dauntless* Training Squadron. HMS *Dauntless*, a hutted camp near Reading, had been the WRNS New Entry Depot since February 1946, where all recruits, both officer and rating, spent their initial four weeks in the Service. Facilities, however, were minimal and could no longer provide the needs of the training programme for the Eighties.

A Superintendent WRNS was appointed to NATO Headquarters, Brussels.

In March 1983 Chief Officer J. Simpson became the first woman to be Deputy Chief Instructor at the Defence ADP Training Centre, Blandford. DADPTC is a Joint Service Establishment specialising in teaching Service and Civil Service personnel in all aspects of computers. Chief Officer Simpson was the first naval representative to hold this post. A number of WRNS ratings work at Blandford and Leading Wren Lee wrote the following in an article on her reactions for the *WRNS Newsletter*.

When I arrived at DADPTC it was not quite what I expected. Somehow I had imagined smart new buildings housing computers with flashing lights, but there was none of this; instead I was faced with a place that looked very similar to *Dauntless*.

WRNS Officer of the Day performs the sunset routine.

WRNS Engineering Officer inspecting the air brakes on the squadron aircraft.

Above: *WRNS helicopter ground crew on deployment at sea in* RFA Engadine *in 1981.*

Below: *The all female watchkeeping team from Flag Officer Sea Training's Operations Room,* HMS Osprey.

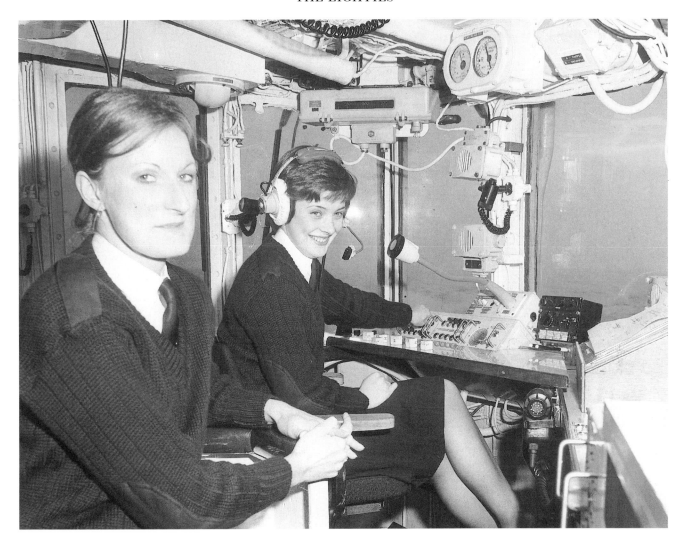

A visit to HMS **Bulwark.**

When I started work I wasn't much use to anyone before I had been on the Fundamentals of Computing course.

In June 1983 Flag Officer Sea Training's Operations Room at HMS *Osprey*, had the first entirely female watchkeeping team.

Second Officer M. Greenaway was the first WRNS officer to work in the Falkland Islands.

An article on my time in the Falklands. . . Obviously I must choose one aspect, but the problem was deciding which one. Should it be the journey south which involved ten days on ss *Uganda*, or life as the only woman in Headquarters British Forces Falkland Islands, which included queueing for meals clutching my very own tin plate, or the day at sea on board the destroyer HMS

Exeter, or the Hercules flight over South Georgia or even the exhilarating experience of being winched down to a submarine? Then there was the day I scrambled over the battlefields of Mount Tumbledown, minefield map in hand, only to be told the following morning that booby traps were still being found in the 'clear' area that I had crossed! There were also the many amusing incidents I encountered while briefing a wide variety of people in an even wider variety of places, the most notable of which was when, on completion of my brief to the two hundred new arrivals on board *Cunard Countess*, I promptly sat on a colleague's lighted cigar and rapidly left the room!

In fact she chose to write about a day trip by Gazelle helicopter which took her to a number of points on East and West Falkland.

On 5 August 1983 Officer Structure Post Command 8288 stated.

Wren Louise Addy received the Princess Anne Novices Jumping Cup in 1980.

Wren Babs Reid in action

That the WRNS should continue as an independent Service headed by a Commandant, and that good career and promotion prospects, and wider job opportunities for WRNS Officers are to be maintained on the basis of existing WRNS Officer structure.

In September there were three 'firsts' for the WRNS. Third Officer Millard became the first WRNS officer to win the Lyddon Shield for the best Supply Officer under training; Wren Wendy Hinton became the first WRNS rating to be formally drafted to the ship's company of a seagoing Royal Naval ship when she spent three weeks aboard HMS *Arrow* as the replacement Navigator's Yeoman; and Third Officer Kahn became the first WRNS officer to win the Rowallan Trophy awarded to the officer under training who displays the greatest leadership potential in his or her first term at Britannia Royal Naval College, Dartmouth.

In December Leading Wren (Regulator) Beryl Worvell was awarded the Brian Welsh Trophy for obtaining the highest marks on a Leading Regulator Qualifying Course during 1983.

On 1 January 1984 the Leadership Course became mandatory for confirmation as Leading Wren.

Leading Wren Regulator Beryl Worvell receiving the Brian Welsh Trophy from Vice-Chief of the Defence Staff (Personnel and Logistics).

Complementing the professional life of naval personnel are the many and varied extra-mural activities which are an essential factor in life. Sport, expeditions, Adventurous Training and charitable pursuits number amongst these. The Women's Services all have creditable records in the full range of recognized sports which, today, include free fall parachuting, Triathlon and hang gliding. Marathon running and support for local and national charities is a feature of naval life. Expeditions involve participation in Joint Service Ventures as well as single Service events, such as the thirteen WRAC and seven Wrens, all stationed in Hong Kong, who joined Exercise Jungle Maiden II

– a ten-day course in jungle survival with the training team in Brunei.

Leading Wren Writer (Pay) Elizabeth Spencer wrote of the experience for the *WRNS Newsletter*.

Gone was the pleasant HMS *Tamar* air conditioning and instead. . . a training programme that sent a few hearts shuddering with remarks such as '0700 – Fitness Training three mile run'. . . lectures in a variety of subjects such as health and jungle navigation, and practical helicopter drills. . . Combats, jungle boots and hat, webbing and Bergen – the real thing was about to begin. In an uncomfortable Army four tonner we made our way to the training area. . . Our hammocks sunk to the ground, our traps caught nothing but our fingers, and our fires lacked a little something. . . practice makes perfect and what we lacked in knowledge was equalled in motivation and an eagerness to

PO Wren Myra Dees jumps to block during the Inter-Service Tournament, which the WRNS won. Behind her, Chief Wren Diz Hadley keeps her eye on the ball.

the chapel in Royal Naval College, Greenwich into the safekeeping of the Reverend Edward Thompson.

The following extracts from the *WRNS Newsletter* are included to show quite clearly the degree to which the WRNS is inextricably woven into the life of the Royal Navy. The first reflects life at Royal Naval Air Station Culdrose whilst the second, at the other end of the spectrum, tells of a Chief Wren on Loan Service with the Royal Brunei Armed Forces.

The WRNS has come a long way since its formation in 1917 and is now more buoyantly alive and moving with the times than it ever was. Certainly here at RNAS Culdrose – the largest WRNS Unit in the country – we have over 200 girls, all equally trained, virtually equally paid and sharing equal responsibility with their male counterparts. . . We have over 20 different WRNS categories and trades represented at Culdrose, many of them totally interchangeable with the men, husband and wife teams co-exist very amicably in a number of departments. . . The biggest single employer is undoubtedly the Supply Department providing personnel, material and catering support for the Front Line Squadrons. . . A high degree of administrative and secretarial expertise is necessary to handle all the Station's correspondence and run the pay accounts of every Serviceman. . . no easy task. In the Communications Department Wrens work 24 hours a day transmitting and distributing signals, operating teleprinters and manning the telephone exchange. There are eleven WRNS officers whose appointments include Public Relations, Squadron Staff Officer, Air Staff Officer, Senior Communications Officer and Air Traffic Controller. . .

First Officer Shepherd teaches meteorology to student aircrew with the invaluable help of the Wren Training Support Assistants who produce all the slides, viewgraphs and films so vital to good instructional technique. . . Fleet Chief Wren Blinston teaches GCE O Level subjects assisted by Petty Officer Wren Education Assistant Thorn and Leading Wren Kitchman who tutor sailors for the naval Maths and English tests.

Out in all weathers on windy hardstandings the Wren Air Mechanics really come into their own. Working around the clock. . . to keep the Sea Kings of 706 and 810 Squadrons at maximum serviceabil-

learn. . . When we had to cross water by precariously balancing across fallen logs, the legs appeared to get a speed wobble and jungle boots certainly were not made for a firm grip. The jungle seemed to enlarge all bugs, ants especially. . . sleeping at night in the jungle is an experience, even though our instructors had joked 'sleep well', for many their eyelids had remained very open and alert. . . It was with mixed feelings that we crouched waiting for the scout helicopter to airlift us to the waiting transport. . . We had all achieved something personal, we had found out and tapped unknown strength and fortitude, whilst retaining that all important sense of humour.

In September 1984 a new category Education and Training Support replaced the Education Assistant and Training Support Assistant categories.

On 28 October 1984 St Mary-le-Strand became the WRNS church. At the dedication service the WRNS Book of Remembrance was transferred from

The Navy still swabs the deck!

*Wren Writer Wendy Cuthberton (left),
Leading Radio Operator Bernie Halsey (centre)
and PO Wren Education Assistant Angie Evans*
*attempt to lift the World Record turkey at the
British Turkey Federation's annual weigh-in,
December 1982.*

Director WRNS (centre) and her staff after the re-gilded weather vane, paid for by the WRNS, had been replaced at St Mary-le-Strand.

ity. Leading Wren AEM Taylor and Wrens AEM Gregory and Morris are three such girls, working with 810 Squadron, who have seen the introduction of Operational Flying Training at sea, and have embarked with the detachments on several occasions in RFA *Engadine*.

Radar Wrens maintain contact with the aircraft on voice circuits, give up-to-date weather reports, plot the progress of the various operations and assist the Air Traffic Controllers. . . A lesser known branch is that of Family Welfare Services – a particularly exacting and diplomatic job demanding special qualities. . . during the height of the Falklands crisis over 1,000 men from Culdrose were down south and Chief Wren Foord did an outstanding job supporting the wives and families waiting at home.

To detail the duties of all the WRNS categories at Culdrose would be exhaustive, but the activities

of the Unit in recent months have included various money raising schemes for local charities and organisations. . . Let it be known that happiness at Culdrose has never been more Wren shaped!
Second Officer S. J. Eagles

The first two months of all Loan Service Personnel's tour of duty in Brunei is taken up by a six-week Basic Malay Language Course, followed by a Jungle Familiarisation Course. I somehow managed to pass the Language Course and prepared myself to go into the jungle. . . We were flown by helicopter and dropped in a clearing. . . From there we walked about one mile and used our parangs (big knifes) to chop a pathway big enough for our party of 30 to follow. With the midday heat, a full pack on my back, full belt order, a parang in one hand and a crate of beer under my arm. . . I think that was the longest mile I've ever walked. . . We were given a demonstration of how to make a basha (bed). . . half an hour later we were all leaping around like Tarzan (me Jane) and chopping down trees in preparation to build our bashas. . .

Tired out Wrens after the completion of their sponsored bed-push in aid of the South Atlantic Fund.

Someone chopped down a tree containing a Hornets nest and we were plagued by the beasts for the next two days. . .

My training period over I started work as Senior NCO in Muara Platoon of the Women's Company . . . not particularly difficult but at times long hours . . . Apart from administration there was the fitness training, sport, parades, welfare and being Duty Officer.   **Chief Wren AEM(M) L. P. Baker**

In January 1985 a Superintendent WRNS was appointed as Chief Staff Officer (Administration) to Commander-in-Chief Naval Home Command.

On 1 June 1985 notice for discharge on marriage was increased to nine months.

On 17 August 1985 Second Officer Jane Bryant was the first member of the WRNS to celebrate her marriage in the WRNS church.

In 1985 Third Officer D. Heesom became the first WRNS student Engineer Officer to undertake the degree course at the Royal Naval Engineering College, Manadon.

She had joined the WRNS in 1983 as a cadet and, after basic training, became a Weapons Analyst.

A Wren Photographer taking aerial shots.

Following an appointment at Royal Naval Air Station, Yeovilton she sat her Admiralty Interview Board in January 1985 and was selected for entry to Dartmouth in May. Her formal request for training as an Engineer Officer was accepted and she started three years study for a B.Eng. degree.

Second Officer D. E. James, a Communications Officer, joined 13 Signal Regiment, a unit attached to the British Army on the Rhine, for a two year appointment in 1985. In her article for the *WRNS Newsletter*, towards the end of her tour, she said:

Within days I discovered that I would not be able to retain my 'blue' identity completely. I was shown the way to the Quartermaster's Stores and kitted out in a rather unbecoming little outfit – combat boots, jacket and trousers, the inevitable set of puttees and worst of all a beret! But what cap badge was I to wear? This little dilemma was solved when Chief Wren at Dartmouth very kindly popped down to the local toy shop, bought a tin of blue airfix paint and painted a Royal Naval officer's cap badge. This badge was needless to say a rarity in Germany and was regarded as a collector's item. I was therefore always careful to hide my beret when I removed it from my head.

On arrival at the Regiment I held a temporary appointment as Troop Commander of the Commcen . . . The Army places great emphasis upon Adventurous Training. My first year I took a group of WRAC to Norway for a week of trekking and canoeing down the fiords and into the coastal area at Kristiansand. . .

This year I thought I would be a little more ambitious. . . and decided Australia offered great potential. . . That is where I am sitting at the moment. . . tropical Cairns, northern Queensland . . . I do have an office! I have enjoyed working with the Army. . . don't ever let it be said that 'Blue and Green should never be seen'.

On 1 October 1985 automatic promotion to Second Officer was introduced, based on seniority as a Third Officer.

In October a Superintendent WRNS became the first WRNS officer to be appointed to a Naval Staff Division as an Assistant Director.

On an exchange program called 'Sea Surge', First Officer M. Farrall was appointed to New Zealand for a four month tour in 1985.

Above: *A Wren working as member of a Buffer's Party.*

Below: *The Falkland Islands Contingent, June 1986.*

One of the first things the naval party (myself, a Warrant Officer, four Chiefs including one Chief Wren, two POs, a Leading Writer and Naval Nurse) learnt about New Zealand is that it is a very long way from anywhere. . . The WRNZNS ceased to exist in 1977 and the women absorbed into the RNZN now comprise about ten per cent of the Service. They hold the same ranks as their male counterparts and wear gold braid and badges. . . As New Entry Training Officer my responsibilities were for the initial training of all entrants, including apprentices. . . and for the administration of most trainees on first professional courses. . . As a Personnel Selection Officer, seeing the 'other side' of training was very useful. I spent more time on the parade ground than ever before. . . and certainly

Wrens in Hong Kong, 1986.

more time doing kit musters than I ever hope to do again.

In January 1986 WRNS personnel started to deploy to the Falkland Islands for four month tours of duty.

Wren Angie Ellis went to the Falkland Islands towards the end of 1987 and wrote 'First Impressions of an F.N.G. (Falklands New Guy)' for the *Newsletter*.

'This is it', I thought, 'I'm off'. I didn't know whether I felt sad or excited as the RAF Tristar took off from Brize Norton. . . Safely back on the ground the hectic part started, trying to find your luggage and the person who was to meet you and show you around. . . the accommodation was better and warmer than I had expected, although the colour of the water and, therefore, that of the sheets we were given, left a lot to be desired! No sooner

had I put down my bags than I was whisked off to be shown where I would be working, which was the Joint Operations Centre. From there I was shown the galley. . . allowed to make my bed before being whisked off to my first Falklands Island party. . . By nine o'clock the jetlag of my 18 hour flight began to creep up on me, so I retired gracefully and fell into my bed.

The next morning the girl I relieved started back on her journey to the UK and I was left to fend for myself. . . During the next two weeks I learnt a great deal. My job was relatively easy to pick up and once I was trained and put into the watch system it meant that I got quite a good amount of time off. The weather was terrible and did nothing but snow, rain and blow a gale. . . At first everybody seemed to get letters except me and they all looked as though they were enjoying themselves. Where was I going wrong? I took wrong turnings for the

A Leading Wren Physical Training Instructor showing junior ratings the ropes.

NAAFI, the gym and the accommodation reception, and spent most of my time totally lost. . .

I am now halfway through my time and finding life at RAF Mount Pleasant really good. I now know plenty of people and the Army, Navy and Air Force personnel mix quite readily. There are lots of good parties and other social gatherings such as quiz nights, and even bingo if you fancy your chances. The sports facilities are excellent and very well used. . . I've paid visits to the Naval ships on patrol off the islands and also the patrol vessels and it's amazing how many people you bump into that you know from UK!

I regularly go flying with 78 Squadron (Chinook helicopters) Search and Rescue Team as a volunteer survivor so that they can practise their hover and winching skills. I have been on a tour of the West Island by Chinook and saw the only substantial clump of trees on the Island! My visit to Sealion Islands was nothing less than fantastic. . . All in all, I am happily enjoying myself and just hope that I can fit everything I want to do into my remaining

time, until I become F.O.G. (Falklands Old Guy).

WRNS personnel have gradually been absorbed into the full range of establishment duties. Petty Officer Wren (QA) Kathy Quinn records one such duty.

'How would you like to be Chief of the Watch on the Main Gate?', my boss asked me on a January morning in 1986. 'Who, me?', I replied. 'Yes, you. They are short of manpower and thought you might like to give it a try.' I really did not know what to expect but on the day that everyone else went on Easter Leave I began my first week of days. I assumed that with Leave everything would be nice and quiet, but this was not to be.

During the next 16 months as CPOOW I was to deal with many day-to-day problems. HMS *Daedalus* is the focal call-out point for Married Quarter defects in the area and many of the calls came from

Director WRNS with Marine Wrens at CTCRM Lympstone, 1986.

young Naval wives on their own needing assistance with leaks in ceilings or fused lights. Other cases were more serious... The hours on watch were very long and tiring, always on the alert and smartly dressed, ensuring that those entering and leaving the establishment were correctly dressed or afforded the proper marks of respect. And, of course, there is living with the uncertainty of not knowing what the next telephone call might bring.

On 3 July 1986 Her Royal Highness The Princess Anne opened the first permanent WRNS exhibition at the Fleet Air Arm Museum.

Exchange appointments with the United States Navy have become a regular practice since the early Seventies. Second Officer C. E. Manley who was at the Officer Candidate School in Newport, Rhode Island, wrote in the *WRNS Newsletter*.

Since I arrived in the USA eighteen months ago I have found that my life has fallen into two parts; work and tourism. When I first arrived at the Officer Candidate School I taught 'personnel

administration' which encompassed everything from US Navy organisation to rules for pay, advancement and leave. . . After five months I became what is known as a Company Officer responsible for about 40 young men and women who wish to become officers in the USN. The majority of them are fresh from university or college but a few have had prior military service or civilian jobs. I spend much of my time instructing, encouraging, counselling, admonishing, inspecting and watching.

In January 1987 a Superintendent was the first WRNS officer to attend the Royal College of Defence Studies course.

A Joint Services Expedition left for East Africa in January with the aim of medical research into the development of Acute Mountain Sickness and to introduce club members to quality climbing at altitude. The expedition comprised twenty men and two women, one of whom was Second Officer M. J. Grimley, Submarine Flotilla Photographic Officer.

We left Heathrow with crates of climbing equipment, medical gear and bulging rucksacks. As the expedition photographer I was carrying dozens of films, all of which had to be opened and scrutinised at the X-Ray desk. I was not popular at the check-out exit! The heat of Nairobi hit us fourteen hours later. . . Leaving the town in a series of battered

A Second Officer WRNS instructing a Weapons Analyst.

local taxis, we headed off for the nearby army camp at Batlsk. . . four days sorting kit and dividing it into (almost) equal loads, and finalising diplomatic clearances. . . We also started the medical tests, recording the results as a base line for the comparative results we were to find on the mountain at different heights, all of which were to change dramatically throughout the five weeks before returning home.

Day five, and we were ready to head for the mountain. . . six hours later. . . at 10,000 feet and already some people were suffering from mild headaches and nausea. We looked more like a bunch of ailing tramps than mountaineers. Even the doctor looked sick!

Next morning had an early start. . . By midday we had reached the edge of the infamous 'vertical bog', the ten mile swamp and tussocks of grass area which divides Mount Kenya from the tropical forests and arid plains of Africa. . . We followed a river running through the hillside to the head of the Teliki Valley, and set down our loads. This was Mackinders Camp, a large clearing on the mountainside where we were to establish base camp for the next three and a half weeks.

Medical tests continued throughout the weeks on the mountain, the most intensive studies being done during the first five days at base camp. Individual reaction to this altitude varied considerably. . .

The ultimate challenge for everyone was to reach the summit of Mount Kenya. My group of four, two climbing pairs, had waited until we felt best acclimatised before we attempted the route. In other words, we could walk twenty paces without having to stop for breath, instead of wheezing after every four steps!. . . By the time we roped up for the last move, the summit was in cloud with ominous dark skies heralding bad weather. We took a few swift photographs and prepared to descend. . . Mission accomplished on Mount Kenya, two days later we packed up the camp, donned rucksacks and walked through the valley, across the 'vertical bog' to the hut, down the long winding track and out of the National Park gates. . .

Two days later we piled in another wagon, heading off in the heat of the sun, to Tanzania. . . we caught sight of Mount Kilimanjaro. . . By sunrise we had gained 2,000 feet, just 2,340 feet more to go, to the crater run and summit glaciers. By midday

Wrens on deployment with Royal Marines in Norway in 1985.

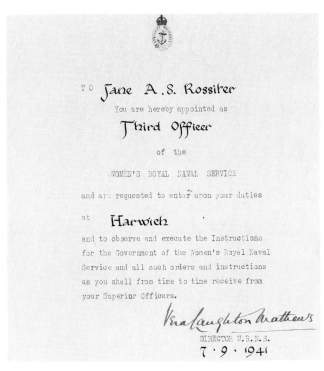

Mrs Rossiter's appointment to Harwich in 1941.

we sat wearily on the 'Roof of Africa', the crater of the largest dormant volcano on that continent. Perched in the snow, above cliffs of solid ice, we looked across the desert plains to the moorland, forests and equatorial zone. So much contrast in so short a distance. Euphoria defeated fatigue as we each felt a sense of achievement. . . It had been a successful expedition.

In February First Officer Caroline Coates was the first WRNS officer to win the top student prize on the Royal Naval Staff Course.

On 2 June 1987 Mrs Jane Rossiter, who was Wren Number 1 in the Second World War died.

Probationary Third Officer J. Neep was the first recipient of The Doris Graham Memorial Prize, at the WRNS OTC Dartmouth, awarded for character and leadership.

Director WRNS opened the sixteenth of 18 NATO Telegraph Automatic Relay Equipments (TARE) at

The Doris Graham Trophy.

H.R.H. The Princess Anne talks to Wrens during her visit to HMS Centurion *in 1987.*

NATO Headquarters, Northwood, on 24 June 1987. It was the first TARE to be manned by an all female staff – in this case Wrens.

In September Ground Crew Wrens were deployed with 899 Naval Air Squadron to Sardinia. Chief Wren AEA(M) Winsor wrote an account of this 'first time' for the *Wrens Newsletter*.

'The list is up for Deci', came the cry. Should we put our names down we wondered; after all we had only ever got as far as the list before, cries of 'no accommodation for Wrens' still ringing in our ears from previous attempts to sample the delights of far-off Sardinia. But being ever optimistic seven female names were scribbled on the list. . .

A movement order appeared and together with fifty male junior and senior rates our names were there in black and white; Chief Wren Angie Winsor, Leading Wren Stacey Clark, Wren Jo Holman, Wren Anne Holden, Wren Rhona Wailes, Leading Wren Trina France-Sargent, Wren Claradette Greenhill and our borrowed nurse ENG Moira Powell. The Wrens were going to Deci at last.

To RAF Lyneham . . . we boarded the Hercules . . . settled down. . . due to the noise level normal conversation became impossible and we were all soon wrapped up in our own thoughts. . . 'Please fasten your seat belts'. . . the heat surged in; at last we were there – Decimomannu. . . Lectures on where to go and where not to go. . . off to our accommodation. . . no time to explore; the real reason for us being here would soon be joining us, our four Sea Harriers.

Back to our 'line' to begin the task of allocating crews for aircraft, unloading ground equipment and spares. Then our 'ops' girls rang through – our aircraft were in circuit, our work had begun. The pilots walked in: 'Had a good flight, Sir?', and more important, 'Is it serviceable?'

Director WRNS visits the WRNS OTC during their Moor Exercise, October 1986. (M.H.F.)

'Then and Now'. Taken at the 1987 WRNS reunion – 70 years since the first formation of the Service. Left to right: Wren Heather Daniels (aged 20), Mrs Ada (Paddy) Worthington (aged 95), Mrs Margaret Parsons (aged 89), Mrs Christina Spendiff (aged 90) and Leading Wren Dee Nelson WRNR (aged 28).

BOOK OF REMEMBRANCE

1914–1919

Elizabeth BEARDSALL *18 December 1918*
Hilda May BOWMAN *24 October 1918*
Margaret Louise CARE *28 October 1918*
Josephine CARR *10 October 1918*
Lucie Emma CLARK *24 November 1918*
Helen Isabella COURT *15 November 1918*
Caroline Jackson DAVIES *26 October 1918*
Harriet Hawkesworth DREWRY *31 October 1918*
Georgina DRYSDALE *23 November 1918*
Charlotte Sophie DUKE *14 August 1918*
Alice May FLANNERY *9 August 1918*
Sarah Ann HALL *14 December 1918*
Bessie Sim HUNTER *15 March 1919*
Lucy Alexander HUNTER *18 November 1918*
Alice KNOWLES *29 November 1918*
Mabel LOCKHART *4 November 1918*
Mabel Caroline PEARSON *4 December 1918*
Lucy READMAN *24 August 1918*
Phyllis Annie SKINNER *5 November 1918*
Trinnette TAYLOR *26 November 1918*
Dorothy Maria WHITE *22 October 1918*
Susan Sophia WILLS *4 October 1918*
Mary WOODRUFF *18 October 1918*

1939–

Chief Officer Kathleen ACKERLEY *31 August 1940*
Wren Rosa May ARMSTRONG *23 November 1940*
Wren Barbara May ALLENDER *31 May 1943*
Petty Officer Wren Evelyn Mary Theodosia
    AITCHISON *6 September 1943*
Wren Jean ASPIN *28 November 1943*
Wren Peggy May ADAMS *4 March 1944*
Leading Wren Joan Margaret ASHBURNER *9 June 1944*
Leading Wren Florence Emily ANDREW *25 January 1945*
Wren Audrey Dolores AYERS *30 April 1945*
Wren Emily ANDREWS *5 June 1946*

Wren Joyce Millicent BENNETT *23 November 1940*
Wren Mona Violet BLACK *23 November 1940*
Wren Winnie BLACKETT *23 November 1940*
Wren Jane BROWN *10 January 1941*

Chief Wren Phyllis BACON *19 August 1941*
Chief Wren Margaret Watmore BARNES
    19 August 1941
Chief Wren Cecilly Monica Bruce BENJAMIN
    19 August 1941
Third Officer Cecilia Mary BLAKE FORSTER
    19 August 1941
Chief Wren Dorothy BONSOR *19 August 1941*
Leading Wren Laura Beatrice BESSANT
    13 January 1942
Chief Wren Blanche Irene BUCKLE *5 July 1942*
First Officer Madeleine Victoria BARCLAY
    1 January 1943
Wren Margaret Elsie BENDALL *26 July 1943*
Third Officer Constance Frances BRITAIN
    30 August 1943
Petty Officer Wren Evelyn Ellett BROWN
    29 October 1943
Wren Hilda BRANTON *6 December 1943*
Wren Ellinor Mary BANTING *3 January 1944*
Wren Anne Colleen BRUEFORD *23 January 1944*
Third Officer Jeanette Lillian BARDEN
    12 February 1944
Wren Hazel Mary BATTEN *12 February 1944*
Wren Marie Elizabeth BREAKELL *12 February 1944*
Wren Jean BARRATT *28 February 1944*
Wren Gertrude Joan BACKLER *4 May 1944*
Leading Wren Rosemary Felicity BOND *15 May 1944*
Leading Wren Margaret Elsie Claire BATCHELOR
    9 June 1944
Wren Diana Sydney BIDDLECOMBE *21 October 1944*
Third Officer Vera Kathleen BROWNE *27 June 1945*
Petty Officer Wren Pamela Eleanor Granville
    BRADSHAW *22 March 1946*
Leading Wren Dorothy May BINDON *10 April 1946*
Wren Joan BURGOYNE *27 August 1949*
Second Officer Sybil Mary BURROWES
    4 November 1954

Wren Vera Margaret COWLE *23 November 1940*
Wren Margaret Marion CLARKE *28 January 1941*
Wren Edith CLEMENTS *10 February 1941*
Wren Freda Coralie COATES *26 April 1941*
Third Officer Margaret Eulalie CHAPPE HALL
    18 August 1941

Chief Wren Madeleine Alice COOPER *19 August 1941*

Wren Gladys Vera May COOPER *19 October 1941*

Leading Wren Ivy Winifred CREIGHTON *13 January 1942*

Wren Yvonne Lilian CAPON *5 April 1942*

Wren Gertrude CANNING *30 June 1942*

Third Officer Rachel Hope CHARLESWORTH
12 September 1942

Wren Mary COWIE *23 October 1942*

Leading Wren Pauline Gladys COOPER *3 March 1943*

Wren Kitty Irene CROCKFORD *12 May 1943*

Leading Wren Mary CAPPER *11 August 1943*

Wren Grace Anne CUMMINGS *15 September 1943*

Wren Annie Dick CAMPBELL *11 January 1944*

Wren Gwendoline CRUDEN *16 January 1944*

Wren Agnes Kyle CARLYLE *12 February 1944*

Second Officer Irene Margaret CLUCAS *11 April 1944*

Wren Anastasia COYLE *30 June 1944*

Wren Joan COULTAS *7 July 1944*

Petty Officer Wren Margaret Mary CANNON
25 August 1944

Wren Elsie Sybil CRAKER *27 December 1944*

Wren Mollie May COVENEY *2 January 1945*

Wren Joyce Muriel Amy CHAPMAN *12 January 1945*

Petty Officer Wren Margaret Auld CORMACK
7 March 1945

Wren Hilda Beryl CORBETT *8 March 1945*

Wren Elizabeth COX *22 March 1945*

Wren Mary COOKE *16 July 1945*

Petty Officer Wren Phillis Anne COLLETT
30 August 1945

Second Officer Eileen Russell CROCKER
1 September 1945

Third Officer Sybil Elizabeth COXON *16 October 1945*

Third Officer Margaret Roseanne COLE
19 January 1946

Petty Officer Wren Alice COPESTAKE *31 January 1946*

Wren Olive Mary CLARKE *22 April 1946*

Wren Iris CARTER *25 March 1947*

Leading Wren Florence Minnie CLARKE
25 November 1947

Wren Margaret COOK *6 June 1948*

Wren Brenda Edith CLEAR *29 August 1950*

Leading Wren Orma Constance Forsyth DUNN
27 August 1940

Wren Alice Maud DESBOROUGH *24 December 1940*

Wren Ivy Marion DAVIES *24 January 1941*

First Officer Margaret Ethne Joan DOBSON
7 July 1941

Leading Wren Anne Archibald DRUMMOND
18 March 1943

Wren Marjorie Elizabeth DORRELL *3 September 1943*

Wren Dorothy Winifred DEBNAM *5 November 1943*

Wren May Sarah Ann DAWSON *27 November 1943*

Leading Wren Winifred Beach DALTON
12 February 1944

Third Officer Cicely Coppard DEAN *12 February 1944*

Wren Joan Ruth DUNCAN *18 June 1944*

Leading Wren Alison Saidee Fitzgerald DALTON
1 February 1945

Leading Wren Edith DAVIES *13 June 1945*

Leading Wren Margaret Dickie DIXEY
13 September 1945

Leading Wren Jean Olwen DIXON *15 December 1946*

Wren Violet Maud DINGLE *31 January 1953*

Wren Maureen EVANS *22 April 1942*

Wren Lily Edith EVERETT *12 May 1943*

Wren Marie ENRIGHT *31 May 1943*

Leading Wren Marjorie Eleanor ELLIS *17 January 1946*

Leading Wren Joan Doreen EMERSON *4 March 1946*

Wren Agnes EGAN *27 August 1946*

Wren Esme Ellen FOWLER *18 February 1940*

Wren Kathleen Battershill FINBOW *5 June 1940*

Wren Margaret Jean FORGIE *12 January 1942*

Wren Mary Chalmers FARQUHARSON
27 March 1943

Leading Wren Gladys FLETCHER *12 February 1944*

Wren Kathleen Elizabeth FIELDING *13 April 1944*

Wren Edith Anne FARMER *18 June 1944*

Wren Ethel Doreen FIELD *22 August 1945*

Wren Cynthia Evelyn FULLER *16 March 1946*

Wren Annie FORSHAW *11 October 1946*

Leading Wren Nellie GREGORY *23 November 1940*

Chief Wren Mary GRANT *19 August 1941*

Chief Officer Charlotte Leila GRITTEN *4 January 1942*

Wren Dorothy Marion GRANT *25 October 1942*

Wren Gladys Gaynor GRIFFITHS *9 May 1944*

Wren Isabel Kathleen GARBUTT *23 January 1945*

Wren Sylvia Heather GIBSON *26 February 1945*

Petty Officer Wren Pauline Mary GOMPERS
27 July 1945

Wren Margaret Hamilton GILLIES *29 November 1945*

Superintendent Ethel Mary GOODENOUGH
10 February 1946

Petty Officer Wren Nora Kathleen GILES *6 April 1946*

Wren Elizabeth Olwen GLADSTONE *3 November 1948*

Wren Doris HOOK *16 June 1941*

Third Officer Isabel Mary Milne HOME *19 August 1941*

Wren Pamela Mary HARVEY *16 May 1942*

Wren Joan Mary Florence HUZZEY *28 January 1943*

Leading Wren Annie Crawford HOSIE
10 February 1943

Leading Wren Joan Mary HUGHES *18 March 1943*

Wren Isobella Scott HYND *24 June 1943*
Wren Diana Vera HARTWELL *14 September 1943*
Wren Rose Barbara HARRIS *9 December 1943*
Leading Wren Elinora HARE *1 January 1944*
Wren Ethel Margaret HUNTER *12 February 1944*
Wren Doris Annetta HOVERD *2 August 1944*
Wren Helen Muirhead HUNTER *15 August 1944*
Leading Wren Dorothy Eunice HARRIS
   12 February 1946
Wren Joyce Evelyn HILEY *3 September 1946*
Second Officer Vera HABERFIELD *19 September 1946*
Petty Officer Wren Olive Alice Enid HAIGH
   2 December 1949
Wren Maureen HORTON *18 September 1954*

Second Officer Nora Patricia INGHAM
   17 September 1945

Third Officer Cecilia Alix Bruce JOY *19 August 1941*
Wren Irene Ethel JONES *23 July 1942*
Petty Officer Wren Mary Lilian JOHNS *1 October 1942*
Second Officer Anne Elizabeth JAGO BROWN
   18 March 1943
Third Officer Thelma Daphne Trench JACKSON
   23 July 1944
Leading Wren Cynthia Patricia Mary JACKSON
   12 August 1944
Wren Catherine Peggy JONES *17 October 1944*
Leading Wren Catherine JOHNSTON *26 December 1944*
Petty Officer Wren Hazel JOHNSON *5 May 1945*
Leading Wren Gladys Mary JENKS *15 February 1947*
Chief Wren Marion Grace JELLETT *21 November 1956*

Leading Wren Evelyn May KIRK *27 April 1941*
Wren Vera Alice Noel KERRISON *7 March 1942*
Wren Aileen Mercy KILBURN *18 March 1943*
Second Officer Marjorie Eileen KENNEDY
   6 January 1945
Third Officer Barbara KING *8 April 1945*
Wren Jean Muir KIRK *26 March 1946*
Wren Angela Sheila KEEPING *19 January 1950*

Wren Margaret Yorath LYNES *21 October 1942*
Wren Dorothy Jean LAWTON *31 May 1943*
Leading Wren Lilian Ada LUCAS *12 September 1943*
Wren Alison LISHMAN *26 May 1944*
First Officer Audrey Lucy LANE *21 July 1944*
Wren Moira Ivy LIVESY *11 September 1944*
Wren Doris LINGARD *15 September 1944*
Wren Joan Elizabeth LOOMES *22 February 1945*
Third Officer Frances Barbara LEWIS *29 July 1945*
Wren Eileen Mary LANGHORN *21 March 1946*
Wren Honor May LAWRENCE *13 June 1948*
Wren Jane Kathleen LAWES *23 February 1949*

Wren Barbara Mary Elizabeth MEDLAND
   22 May 1941
Wren Jessie Mary MICKLEBURGH *29 June 1941*
Third Officer Victoria Constance McLAREN
   19 August 1941
Third Officer Florence MACPHERSON *19 August 1941*
Third Officer Kathleen MILLER *19 August 1941*
Wren Mary Cavell MACLACHLAN *23 January 1942*
Chief Wren Iris MARLOW *8 February 1942*
Wren Jean McBURNEY *19 April 1942*
Leading Wren Joan Catherine MARCH
   4 September 1942
Leading Wren Marion Grace MELHUISH
   28 January 1943
Wren Joan Peggy MORSLEY *7 February 1943*
Wren Annie Elizabeth MACORMICK *31 May 1943*
Wren Lilian MARTIN *26 October 1943*
Wren Elizabeth Hilda MAIR *27 November 1943*
Wren Betty Grace MAYES *1 May 1944*
Petty Officer Wren Marion McGLINCHY *8 June 1944*
Wren Betty MEYNELL *7 July 1944*
Petty Officer Wren Kathleen Janet Mary MALINS
   5 August 1944
Wren Jeannette MARTIN *6 August 1944*
Leading Wren Eva MONKS *26 October 1944*
Wren Edith Maud MARTIN *21 February 1945*
Third Officer Mary Macdougall MOLLOY
   16 March 1945
Wren Jean McCULLOCH *3 May 1945*
Second Officer Betty Helen Whitson MACAULAY
   13 June 1945
Leading Wren Phyllis Olga MORRIS *19 June 1945*
Wren Margaret Edith MOXLEY *25 September 1945*
Wren Annie MACKINTOSH *1 October 1945*
Chief Officer Dorothy Mary MACKENZIE
   7 October 1945
Wren Barbara Joyce MASTERS *30 October 1945*
Leading Wren Anne Gordon MACFARLANE
   27 December 1945
Wren Helen Adair McQUAKER *31 December 1945*
Wren Dorothy Edith MUSK *11 April 1946*
Leading Wren Catherine Mary MACDONALD
   2 July 1946
Leading Wren Mary Ellen MADDERN *9 July 1947*
Leading Wren Jean Margaret MURDOCH
   21 May 1953
Wren Jean McWHIRTER *19 December 1954*

Wren Beryl Melita NORTHFIELD *23 November 1940*
Chief Wren Mildred Georgina NORMAN
   19 August 1941
Third Officer Nesta Margaret NEWMARCH
   30 August 1943
Wren Doris Evelyn NORMAN *20 November 1943*

Wren Aileen Audrey Buchanan NICKSON
12 February 1944
Wren Beatrice Marjorie NYE *12 February 1944*

Second Officer Christina Emma OGLE *19 August 1941*
Wren Margaret Mary O'NEILL *17 October 1945*
Wren Doreen Charlotte ORD *4 January 1951*

Leading Wren Hilda PEARSON *23 November 1940*
Wren Annie PYNE *30 April 1942*
Wren Hilda Marion PARK *25 June 1942*
Third Officer Pamela Ida PRICHARD *12 July 1942*
Petty Officer Wren Florence Elizabeth PLACE
30 October 1942
Leading Wren Violet Bessie POWELL *18 March 1943*
Wren Olive Doreen PETT *27 June 1943*
Wren Anita Ella PULLEY *16 August 1944*
Leading Wren Peggy Irene PERKINS *4 April 1945*
Third Officer Monica Joan PEARSON *6 October 1945*
Wren Ruby POTTER *6 August 1946*
Third Officer Margaret Elizabeth PODMORE
22 June 1947
Wren Eileen Sylvia PRENTICE *31 January 1953*
Wren Barbara Luetchford PALMER *15 July 1956*
Wren Norah Kathleen REYNOLDS *23 November 1940*
Wren Sybil Doreen REDDING *7 March 1941*
Third Officer Josephine Caldwell REITH *19 August 1941*
Wren Mary Edith Lydia ROBINSON *30 January 1942*
Wren Muriel Alfreda RAINTON *18 March 1943*
Wren Ellen Elizabeth REGAN *18 March 1943*
Wren Violet ROBERTSON *16 September 1943*
Leading Wren Eve Annette RAINEY *15 October 1943*
Pro. Wren Audrey Constance READ *2 November 1943*
Third Officer Marion Carson ROBINSON
12 February 1944
Wren Daphne Jean REVELL *15 October 1945*
Wren Edna May RAWSON *23 September 1946*
Leading Wren Irene Marion ROBERTS *22 February 1947*
Wren Simone Yolande RABET *21 October 1947*
Petty Officer Wren Florence Powell RODGERS
7 June 1949

Wren Gladys Irene STEPHENS *2 January 1941*
Wren Violet Ellen Elizabeth SMITH *7 July 1941*
Chief Wren Elsie Elizabeth SHEPHERD *19 August 1941*
Chief Wren Catherine Johnston SLAVEN *19 August 1941*
Chief Wren Beatrice Mabel SMITH *19 August 1941*
Wren Hazel Kathleen SANDERSON *11 September 1941*
Wren Jane Ann SHEPHERD *20 January 1942*
Wren Eileen SHOTTON *8 August 1942*
Wren Catherine Peebles SMITH *15 February 1943*
Wren Catherine STARKEY *28 August 1943*
Wren Jean Young SIME *23 September 1943*
Wren Marjorie SMITH *21 January 1944*

Leading Wren Heather Mowbray SMAIL
12 February 1944
Wren Audrey Hilda STAFFORD *12 February 1944*
Wren Deborah SHUTE *25 February 1944*
Wren Ellen SAUNDERS *23 August 1944*
Third Officer Brenda Quinlan STAFFORD
28 December 1944
Third Officer Daphne Lucy Regina SWALLOW
1 January 1945
Wren Dorothy Joyce STONE *5 April 1945*
Wren Margery Amy STOCKEN *9 April 1945*
Petty Officer Wren Isobel Florence SQUIRES
27 July 1945
Wren Rachel SCOTT *27 August 1945*
Wren Violet Rose SOUTH *29 October 1945*
Wren Sylvia SANDERS *11 December 1945*
Wren Jean Mary SHOOTER *4 January 1946*
Wren Elizabeth Muriel SUMMERS *3 March 1946*
Second Officer Evelyn Monica SANER *29 March 1946*
Wren Nancy Esme SCARFE *22 October 1946*
Wren Iris Betty STILLMAN *24 April 1947*
Wren Lorna May SRODZINSKI *22 September 1947*
Leading Wren Elizabeth Mary STEVENS *26 January 1948*
Wren Dorothy Mary STONE *10 June 1948*
Wren Eileen Mary STALLWORTHY *30 August 1948*

Leading Wren Pamela Annette TANSLEY *7 July 1942*
Wren Elsie THOMPSON *17 August 1942*
Wren Margaret Mary Iris THOMPSON *13 January 1943*
Wren Rita Mary Rose TURNER *5 April 1943*
Margaret Ponton TODD *12 February 1944*
Third Officer Olivia Ann TREVOR *4 July 1945*

Wren Elizabeth VYNER *3 June 1942*
Leading Wren Helen Morag Jean VALENTINE
12 February 1944

Wren Dorothy Margaret WARDELL *23 November 1940*
Chief Wren Ellen Jessie WATERS *19 August 1941*
Chief Wren Rosalie WELLS *19 August 1941*
Petty Officer Wren Ellen Victoria WHITTALL
18 September 1942
Leading Wren Dorothy Downes WILKINS
26 October 1942
Petty Officer Wren Elizabeth Ann WALDEN *5 April 1943*
Wren Mary Joan WILD *13 September 1943*
Wren Catherine WILLIAMSON *3 January 1944*
Wren Betty Ramsey WHITE *12 February 1944*
Leading Wren Pamela Irene WYLLIE *12 February 1944*
Wren Patricia Mary WELLESLEY *28 August 1944*
Leading Wren Evelyn Eileen WILLIAMS *21 May 1945*
Leading Wren Joan Mary WILSON *18 August 1947*
Wren (W.R.N.V.R.) Elizabeth Joan WRIGHT
8 September 1953

Wren Pamela Francis YOUNGMAN *12 February 1944*

Names which have been added since the Book was dedicated on 14 March 1959

Wren Joy Angela SMITH *10 February 1959*
Wren May Beaumont CARTER *23 January 1940*
Chief Wren Marie Isabella McLACHLAN
27 September 1959
Wren Patricia Mary BUCKARD *5 March 1960*
Petty Officer Wren Kathleen May JOHNSTON
3 June 1960
Wren Doreen WOODS *11 October 1960*
Wren Margaret Mackinnon CLARK *16 February 1961*
Chief Wren Stella Violet PEGLER-SMITH *14 July 1961*
Wren Margaret Lillian LUKE *29 June 1962*
Wren Maureen Ruby LANCASTER *21 October 1963*
Wren Christine Mary HUNT *24 February 1966*
Petty Officer Wren Helen Margaret RUSK *5 April 1966*
Leading Wren Celia Elizabeth DODSON *29 May 1966*
Wren Helen Sandra CRAIGIE *31 July 1967*
Leading Wren Betty LEWIS *5 November 1967*
Wren Carol Dorothy HOWE *1 June 1969*
Chief Officer Jean Sutherland RAE OBE *30 September 1969*

Leading Wren Margaret Anne WELLING
11 November 1970
Wren Valerie CARR *10 January 1972*
Wren Irene GIBBONS *10 January 1972*
Third Officer Elisabeth D. PRICE *11 March 1972*
Wren Diane PHILPOTT *30 May 1972*
Chief Wren Joan Enid PAIN *2 July 1972*
Petty Officer Wren Christina M.D. MATTHEWS
15 February 1974
Wren Sheila Margaret HORNE *9 October 1974*
Second Officer Patricia Kay JOLLY *2 November 1974*
Leading Wren Annie Mary BYRNE *29 August 1975*
Leading Wren Tina Janet CHILDS *1 November 1975*
Wren Karen Francesca KINGDON *30 November 1975*
Chief Wren Eileen Arnold EVANS *22 April 1976*
Third Officer Christine JEFFREY *13 August 1976*
Leading Wren Carol Ann BARKER *18 February 1977*
Leading Wren Alison de LOOZE *2 June 1977*
Wren Janet Beryl MASSINGHAM *6 February 1979*
Leading Wren Jacqueline MORRIS *1 February 1980*
Wren Katherine Mary Verdon HARRIS *9 July 1981*
Wren Susan Catherine KEANEY *18 December 1981*
Wren Alexandra Helen TALBOT *3 August 1984*
Chief Wren Irene MURPHY *1 February 1985*

BRANCHES & CATEGORIES

The first two lists give some idea of the variety of tasks WRNS officers and ratings were concerned in during World War I.

Between 1939 and 1943 there was a rapid build up in the variety of work WRNS officers and ratings were called upon to perform. With time these were resolved into named specializations and by the end of 1943 the following were established.

Officers' Branches

Accountant
Administration
Anti-Gas
Assistant Paymasters
Cypher
Fleet Mail
Intelligence
Quarters
Secretarial
Signaller

Ratings' Categories

Anti-Submarine Wire Net Constructors
Book-Keepers
Clerks
Cooks
Depth Charge Primers
Despatch Riders
Fitters
Gyroscope Adjusters
Laundress
Mine Cleaners
Motor Drivers
Porters in Victualling Stores
Respirator Repairer
Sailmakers
Searchlight Lamps and Hydrophone Maintenance and Repair
Stores
Telegraphists
Telephonists
Turners
Waitresses
Wireless Operators

Officers' Branches

Accountant	Flying Control
Administrative	Gunnery (Experimental)
Air Engineer	Instructional Film
Air Maintenance	Intelligence
Air Radio	Mess Caterer
Aircraft Recognition	Meteorology
Amenities Liaison	NCS Boarding & Routing
AA Target	Orthoptist
ARP	Personnel Selection
Armament Stores	Physical Trainer
Boarding	Plotter
Bomb Range Marker	Quarters
Boom Defence Control	Safety Equipment
CB	Secretarial
Censor	Signal
Cine Gun Assessor	Special Duties
Classifier	Sub-Accountant
Cypher	Submarine Attack
DEMS Inspector	Assessor
Drafting	Teleprinter
Education	Topographical
Fighter Direction	Torpedo Attack Assessor
Film Interpreter	Vision Testers
Fleet Mail	Welfare

WRNS Ratings

AA Target	Automatic Morse
Administrative	Transcriber
Anti-Gas	Boat's Crew
Air Mechanic	Boat Driver
(L)(A)(E)(O)	Boom Defence
Air Synthetic Trainer	Bomb Range Marker

Book Corrector
Chart Corrector
Cine Gun Assessor
Cinema Operator
Classifier
Coder
Cook (O) (S)
Courier
Degaussing Recorder
Despatch Rider
Drawing Duties
Fabric Worker
Gardener
General Duties
Gunnery Control
Gunnery (Experimental)
Hairdresser
Laundry Maid
Mail Clerk
Mess Caterer
MQ
Maintenance
Maintenance (Air)
Messenger
Meteorological
Minewatching
M/T Driver
Net Defence
Parachute Packer
Photographer
Photographic Assistant
Plotter
Printer
QO (A)

QO (LC)
QO (CO)
Quarters Assistant
Radio Mechanic
Recruiting Assistant
Radar Operator
R/T Operator
SDO Watchkeeper
Ship Mechanic (LC)
Shorthand/Typist
Signal Exercise Corrector
Special Duties (Linguist)
Special Operators
Steward (G) (O)
Supply (Clothing)
Supply (Naval Stores)
Supply (Victualling)
Supply (FAA Stores)
Submarine Attack
   Teacher
Switchboard Operator
Tailoress
Topographical
Torpedo
T/P Operator
Typist
Vision Tester
Visual Signaller
Writer (General)
Writer (Pay)
Writer (Pay) (DEMS)
Writer (RM)
W/T

By 1947 all the categories which were peculiar to wartime had become redundant, and the shape of future needs were starting to evolve. WRNS ratings were employed as follows:

Aircraft Direction
Air Stores
Cinema Operator
Clothing
Cook (O)
Cook (S)
EVT Instructor
Hairdresser
Mess Caterer
Meteorological
M/T Driver
Naval Airwoman
Quarters Assistant

Range Assessor
Regulating
Steward (G)
Steward (O)
Switchboard Operator
Tailoress
Telegraphist
Victualling
Welfare Worker
Wren Radio (AR)(AW)
Wren (Signal)
Writer (General)(Pay)
Writer (Shorthand)

By the middle of the Sixties the Dental Hygienist and Dental Surgery Assistant had arrived and the Sick Berth Attendant had come and gone; Education Assistant had replaced EVT; the term Telegraphist and Wren Signals had been replaced by Radio Operator and Radio Operator (M); the Naval Airwoman had become Air Mechanic (AE), (A) or (E); Radio Electrical (Air) had replaced Wren (Radio) (AR) and (AW), and Air Direction had disappeared. Range Assessors had been superseded by Weapon Analysts and the Radar Plotter had reappeared.

As the Seventies developed categories continued to evolve in line with the changing face of the Royal Navy. The Hairdresser and Mess Caterer disappeared; Air Fitter, Radio Electrical (Air), and Radio Supervisor were added and the sub-specializations in the Stores were amalgamated to form the Stores Accountant category.

Naval organization of Branches and Categories today are grouped as follows:

Officers' Branches

Executive, Administrative and Operational Branch
Air Traffic Control
Communications
Fleet Analysis
Photographic Interpretation
Personnel Selection
Other tasks:
   Careers
   Computer and ADP
   Intelligence
   Operations
   Public Relations
   Staff
   Training

Engineering
Air Engineer
Weapons Electrical Engineer

Instructor
Teaching
Meteorological
Oceanographic
ADP
Educational Technology and Training Management

Supply and Secretariat
Secretarial
Stores
Catering
Cash

WRNS Ratings' Categories

Operations
Weapon Analyst
Radar
Radio Operator

Supply and Secretariat
Writer
Stores Accountant
Cook
Steward

Dental
Dental Surgery Assistant
Dental Hygienist

Fleet Air Arm
Air Engineering Mechanic
Meteorological Observer
Telephonist
Motor Transport Driver
Quarters Assistant
Education and Training Support
Regulator
Family Services
Physical Trainer
Photographer

SHIPS & ESTABLISHMENTS

Since the first formation of the Service, Wrens have been employed in a wide variety of locations. During the Second World War naval establishments proliferated, to disappear just as quickly once peace had been restored. The following list has been garnered from Service records and information received from ex-Wrens. It is by no means complete but does, I believe, give a fair indication of the mobility achieved by the Service in response to the needs of the Royal Navy.

Afrikander	Simonstown, SA
Aggressive	Seaford, Essex
Ajax	
Allenby	
Almanzora	Troopship
Ambrose	Dundee, Angus
Anchor	
ANCXF	Normandy, France
Anderley	
Anderson	Colombo W/T, Ceylon
Andes	Troopship
Appledore	Appledore, Devon
	Ilfracombe, Devon
Ariel	Warrington, Lancs
	Worthy Down, Sussex
Arkella	Boston, Lincs
Arley	
Armadillo	Glenfinart, Argyll
ASE	Haslemere, Surrey
Assegai	Durban (Transit Camp), SA
Athlone Castle	Troopship
Attack	Portland, Dorset
Aurangi	UK to Durban April 1943
Austin Bay	
Avonmouth	Bristol
Bacchante	Rosyth, Fife
	RNAH Newmacher, Aberdeen
	Peterhead, Aberdeen
	Banff
	RNAH Kingseat
Badger	Dovercourt, Essex
Bambara	
RNAH Barrow Gurney	Bristol
Beaver	Immingham, Lincs

SHIPS AND ESTABLISHMENTS

Ship/Establishment	Location
	Hull, Yorks
	Headingley, Leeds
Bee	Holyhead, Anglesey
Beehive	Felixstowe, Suffolk
	Deal, Kent
Beer Head	
Belfast Castle	
Bherunda	Colombo, Ceylon
Birnbeck	Weston Pier, Somerset
Black Bat	
Blackcap	RNAS Grappenhall,
	Hayes
	Stretton, Lancs
	Paull, East Yorks
	Warrington, Lancs
Boscawen	Lyme Regis, Dorset
	Poole, Dorset
	Portland, Dorset
	RNAH Sherborne, Dorset
	Weymouth, Dorset
Bougie	
Braganza	
Bristol	Bristol, Avon
Britannia	Chester, Cheshire
	Dartmouth, Devon
Brontosaurus	Dunoon, Argyll
RNSC *Brussels*	
Bunting	Ipswich, Suffolk
Byrsa	
Cabbala	Lowton St Mary, Lancs
Cabot	Cardiff
	Wetherby, Yorks
	Bristol
Caledonia	Rosyth, Fife
	Oban, Argyll
Calliope	Newcastle, Northumberland
	Tynemouth, Northumberland
	North Shields, Northumberland
Caroline	Bangor, NI
	Belfast
Carrick	Greenock, Renfrew
Ceres	Wetherby, Yorks
Chittral	Troopship
Chrysanthemum	
Cicala	Kingswear, Devon
	Coastal Forces I
Claverhouse	Edinburgh
Clio	Barrow-in-Furness, Lancs
Cochrane	Rosyth, Fife
	Gullane, East Lothian

Ship/Establishment	Location
	Dunfermline, Fife
	Methil, Fife
	North Queensferry
Collingwood	Portsmouth, Hants
Condor	Arbroath, Angus
Conte	Combined Ops Naval Training, Dartmouth
Copra	London
	Largs, Ayr
Cormorant	Gibraltar
Cressy	Dundee, Angus
Cricket	Bursledon, Hants
	Crossaig Bombing Range
Curlew	Dunoon, Argyll
	Buthkollidar
Cyclops	Rothesay, Bute
Daedalus	Newcastle under Lyme, Staffs
	Bedhampton, Hants
	Defford, Worcs
	Lee-on-Solent, Hants
Dalinda	
Dartmouth	Dittisham, Devon
	Dartmouth, Devon
	Brixham, Devon
	Salcombe, Devon
	Exmouth, Devon
	Seaton, Cornwall
Dauntless	Burghfield, Berks
Defender	
Defiance	Plymouth
Demetrius	Wetherby, Yorks
Denton	
Dido	Parkstone, Harwich
Dipper	Henstridge, Somerset
Dinosaur	Troon, Ayr
Diver	
Dolphin	Portsmouth
Dominion Monarch	Troopship
Dorlin	Acharacle, Argyll
Dragonfly (ex *Northney*)	Hayling Is., Hants
Drake	Devonport, Devon
	Instow, Devon
	RNAH Muristowe
	RNH Stonehouse
	Avonmouth, Glos
	Barrow Gurney, Bristol
	Bodmin, Cornwall
	Henstridge, Somerset
	Wembury, Devon
	Egg Buckland
Dreel Castle	Falmouth, Cornwall
Dryad	Southwick, Hants

153

Duke	Malvern, Worcs	*Forward*	Newhaven, Sussex
Dundonald	Troon, Ayr Combined		Seaford, Sussex
	Ops Trng Centre	FOSNO	Schleswig-Holstein,
RNAS Dunino			Germany
Dunluce Castle	Lyness – Depot Ship	*Fox*	Lerwick, Shetland Is.
RHAH Durdham Down	Bristol	*Fullerton*	
Eaglet	Birkenhead	*Fulmar*	RNAS Lossiemouth,
	Liverpool		Moray
	Malpas, Cheshire	*Gadwall*	RNAS Belfast, NI
	Blundellsands, Lancs	*Gamecock*	Nuneaton, Warwicks
		Ganges	Shotley, Suffolk
RNAS Eastcote		*Gannet*	RNAS Eglinton, NI
Edinburgh	Troopship		RNAS Prestwick, Ayr
Effingham	Royal Naval College,		
	Dartmouth	*Garrick*	Glasgow
Egmont	Malta – Depot Ship	*Garruda*	Coimbatore, India
Elfin	Blyth, Northumberland	*Gayhurst*	
Empire Woodlark	Troopship	*Glendower*	Pwhelli, Caernarfon
Empress of Canada	Troopship	*Glenholt*	
Empress of Japan (renamed		*Glynn*	
Empress of Scotland)	Troopship	*Gnu*	Cape Town, SA
Euphrates	Basra, Iraq	*Godwit*	RNAS Hinstock, Salop
Europa	Lowestoft, Suffolk	*Goldcrest*	RNAS Dale, Pembs
	Wrentham, Suffolk		RNAS Brawdy
	Gibraltar	*Golden Hind*	Sydney, Melbourne and
Excalibur	Stoke-on-Trent, Staffs		Brisbane, Australia
Excellent	Bognor, Sussex	*Gorleston*	
	Southsea, Hants	*Gosling IV*	Warrington, Lancs
	Petersfield, Hants	*Grasshopper*	Lyme Regis, Dorset
Falcon	RNAS Halfar, Malta		Portland, Dorset
Faraway			Weymouth, Dorset
Ferret	Londonderry, NI	*Gregoli*	Malta
Fervent	Broadstairs, Kent	*Greenwich*	Royal Naval College
Fieldfare	Inverness	RNAS Grimsetter	Orkney
Firth		*Gunner*	Granton, Lothian
Fisgard	Torpoint, Cornwall	*Haig*	Dover, Kent
Fledgling	Mill Meece, Staffs	*Halcyon*	Lowestoft, Suffolk
Flora	RNAH Invergordon	*Hannibal*	Algiers
	Inverness	*Harrier*	RNAS Kete, Pembs
Flowerdown	Winchester, Hants	RNAS Hatston	
Flycatcher	Middle Wallop, Hants	*Hawke*	
Foliot	Keyham, Plymouth	*Headingley*	Leeds, Yorks
	Calstock, Devon	*Helder*	Clacton-on-Sea, Essex
	Saltash, Cornwall		St Osyth, Essex
	Tamerton Foliot, Devon	*Helicon*	Ault Bea, Ross & Crom
	Roborough, Devon		Sydney, Australia
Foral	Southwold, Suffolk	RNH Herne Bay	
Force Pluto		*Heron*	RNAS Charlton
Forte	Coverack, Cornwall		Horethorn
	Falmouth, Cornwall		Teignmouth, Devon
Forth	Ardnaman, Argyll		Yeovilton, Somerset
Fortitude	Ardrossan, Ayr	*Highflyer*	Trincomalee, Ceylon
	Lamlash, Ayr	*Hong Siang*	Troopship
Forton	Forton, Hants	*Hopetown*	Port Edgar
Fort Wellington		*Hornbill*	Culham, Oxon
		Hornet	Gosport, Hants

	Shepton Mallet, Somerset
Humming Bird	
Ile de France	Troopship
Indomitable	
Indus	Plymouth
Irrvent	
Irwell	
Jackdaw	RNAS Crail, Fife
James Cook	Tighnabruich, Argyll
Karagola	Troopship
Keobe	
Keron	
Kestrel	RNAS Worthy Down, Hants
Khedive Ismail	Two Wrens saved after being torpedoed
Killarney	Rosyth, Fife
King Alfred	Hove, Sussex Shoreham-by-Sea, Sussex
Kongoni	Durban, SA
Kranji	Singapore W/T Station
Landrail	Campbeltown, Inverness RNAS Machrihanish Skipness, Argyll
Lanka	Colombo, Ceylon
La Pampa	Troopship
Leigh	Southen, Essex
Leyard	
Lizard	Hove, Sussex
Lochailort	Inverailort, Inverness Lochailort, Inverness
Lochinvar	Port Edgar, West Lothian Granton, Edinburgh
Louisberg	Roseneath
Lucifer	Swansea, Glamorgan Barry, Glamorgan Cardiff
Lynx	Folkestone, Kent Kearnsey Kingsgate Dover
RNAS *Lympne*	Dover (later Newcastle under Lyme)
Lyness	Scapa Flow, Orkney
Macaw	Bootle, Cumberland
Maerira	
Malagas	RNAS Wingfield, Cape Town
Manatee	Yarmouth, IOW
Mantis	
Marlborough	Eastbourne, Sussex
Marshal Soult	Portsmouth, Hants

Marshall	Beirut, Lebanon
Marston	
Martello	Lowestoft, Suffolk
Martial	
Mastodon	Beaulieu, Hants
Mauretania	Troopship
Mauritius	Mauritius
Medfire	
Medina	IOW
Medway	S/M Depot Ship
Mentor	Stornoway, Isle of Lewis
Mercury	Basing Park, Bath Hambledon, Hants Haslemere, Surrey Droxford East Meon Leydene
Merganser	
Merlin	Survey Boat, Plymouth Aberdour, Rosyth, Fife RNAS Donibristle, Fife Drem, East Lothian RNAS Evanton, Ross & Crom.
Mersey	Liverpool
Midge	Great Yarmouth, Norfolk
Mill Hill	WRNS Depot, London
Minos	Lowestoft, Suffolk
RNHQ Minden	Germany
Monarch of Bermuda	Troopship
Monck	Port Glasgow, Largs
Monea	
Monster	Rosyth, Fife Inverness
Moreta	Beirut, Lebanon
Mount Stewart	Teignmouth, Devon
Myloden	Southwold, Suffolk Lowestoft, Suffolk Southend, Argyll
NCS	Brightlingsea, Essex
Nemo	Faslane, Dumbartonshire
Neptune	Newhaven, Sussex
Newt	
NHQ & Contraband Control Station	Falmouth, Cornwall
Nighthawk	
Nightingale	Launch in Plymouth
Nightjar	RNAS Inskip, Preston, Lancs Cambeltown, Argyll
Nimrod	Alexandria, Egypt Cairo, Egypt
Nile	
Northney	Hayling Island, Hants Havant, Hants

Nuthatch	RNAS Anthorn	Shrewsbury, Salop
Oban	Naval Base (WWI)	Elton, Berks
Odyssey		Bristol DEMS Pay
Onyx	Torbay, Devon (WWI)	Office
Orion	Troopship	Windsor, Berks
Orlando	Greenock, Renfrew	London
	Helensburgh,	Harwich, Suffolk
	Dumbarton	
	RNAH Kilmacolm,	
	Rewfrew	
Orontes	Troopship	
Orrel Hey		
Osborne	Cowes, IOW	
	Culver Cliff, IOW	
	Ryde, IOW	
	Seaview, IOW	
Osiris	SNOME Egypt 51/53	
Osprey (ex *Attack*)	Fairlie, Ayr (WWI)	
	Portland, Dorset	
Owl	RNAS Fearn, Evanton	
Pactolus		
Paragon	Middlesbrough, Yorks	
	West Hartlepool, Co.	
	Durham	
	Scarborough, Yorks	
Paris	MTB Training & Depot	
	Ship, Devonport	
Pasteur	Troopship	
Peewit	RNAS Easthaven, Angus	
Pekin	Grimsby, Lincs (WWI)	
Pembroke	Chatham, Kent	
	Northfleet, Kent	
	Rochester, Kent	
	Rainham, Essex	
	Strood, Kent	
	Hereford	
	Mill Hill, London	
	Oxford	
	East Tilbury, Essex	
	Brompton, Kent	
	Gillingham, Kent	
	Staines, Middlesex	
	Bletchley, Bucks	
	Stanmore, Middlesex	
Pendragon	Southsea, Hants	
Peregrine		
Philante	Sloop HQ Cinc Western	
	Approaches	
Phoenicia	Lyness and Orkney	
Phoenix	Egypt	
Pomona	Lyness and Orkney	
Porcupine II	Fort Southwick	
Powerful	Plymouth (WWI)	
President	Calmore, Hants	

Princess Irene	Berlin, Germany
Princess Kathleen	Port Tewfik, during
	evacuation from
	Alexandria, 1942
Prometheus	Alexandria, Egypt
Proserpine	Lyness, Orkney
	Thurso, Caithness
	Hoy, Orkney
Pyramus	Kirkwall, Orkney
Quebec	Inverary, Argyll
Queen	Escort Carrier
Queen Charlotte	Ainsdale-on-Sea, Lancs
Queen Elizabeth II	Troopship
Queen Mary	Churchill's visit to USA
Racer	Larne NI
Raleigh	Torpoint, Cornwall
Ramsgate	Coastal Forces
RNAS Rattray	
Ravager	Rothesay, Bute
Raven	Eastleigh, Hants
Redhouse	
Reina del Pacifico	Troopship
Ringtail	RNAS Burscough, Lincs
Robertson	Sandwich, Kent
Robias	
Robin	
Rodent	Liss, Hants
Roedean	
RNAS Ronaldsway	
Rooke	Donibristle
	Gibraltar
Rosemarkie	Rosemarkie, Ross
	Fortrose
Roseneath	Helensburgh, Dumbarton
Royal Alfred	Kiel, Germany
Royal Arthur	Skegness, Lincs (Now –
	Corsham, Wilts)
Royal Charles	Calais (WWII)
Royal Harold	Plon, Germany
Royal Marines	Deal, Kent
	Eastney, Hants
	Gosport, Hants
	Ilfracombe, Devon
	Liphook, Hants
	Penshurst, Kent
	Poole, Dorset
	Thurlestone, Devon

Rusherther	Lympstone, Devon
	Arbroath, Angus
SACSEA HQ	Kandy, Ceylon
STE *Southampton*	Sherbrooke
Safeguard	Southampton, Hants
St Andrew	
St Angelo	Malta
St Barbara	Bognor, Sussex
St Christopher	Fort William, Inverness
St Dunstans	
St George	Douglas, IOM
St Lukes	
St Matthew	Burnham, Essex
St Vincent	Portsmouth, Hants
	London
Saker	Washington, USA
	New York
Salcombe	Exmouth, Devon
Samaria	Troopship
Sanderling	RNAS Abbotsinch,
	Renfrew
Sandhurst	Cairaryan, Strathclyde
Sandpiper	Abbotsinch
Scotia	Ayr
RNAH *Seaforth*	Plymouth
Seahawk	RNAS Culdrose, Helston
	Arnashaig, Argyll
Sea Serpent	Bracklesham, Sussex
Sheba	Aden
Shrapnel	Bournemouth, Hants
	Glasgow
	Southampton, Hants
	Calshot, Hants
Shrike	
Sinclair HC	Troopship
Skirmisher	Milford Haven, Pembs
	Pembroke
Siskin	
RNAH *Southampton*	
Sparrow Hawk	RNAS Hatston, Orkney
Spartiate	RNAH Kilmacolm,
	Renfrew
	Tulliechewan Castle
	Cambusland, Lanark
Speaker	'Woolworth Carrier' –
	Troopship
Squid	Southampton, Hants
Stag	Fayid, Egypt
Stirling Castle	Troopship
Stopford	Bo'ness, West Lothian
	Linlithgow, West
	Lothian
Strathmore	Troopship

Strike	Londonderry, NI
	RNAS Maydown
SEAC	Delhi, India
Tamar	Hong Kong
Tamaron	Troopship
RNAS Tambaram	
Tamoroa	Troopship
Tana	Kilindini, Kenya
Tennyson	Exmouth, Devon
Tern	RNAS Twatt, Orkney
Terror	Singapore
Titania	Belfast, NI
	Londonderry, NI
Torch	
Tormentor	Warsash, Hants
Trelawney	Aultbea, Ross
	Duncraig, Ross
Triphibian II	Harrogate, Yorks
Troon	Flag Officer Carrier
	Training
Turtle	Poole, Dorset
	Dorchester, Oxon
Ubiquity	Granton, Edinburgh
Ukassa	RNAS Katukurunda,
	Ceylon
Unicorn	Dundee, Angus
Urley	
Usaka	
Usley	
Valkyrie	
Varbel	Rothesay, Bute
Vectis	Cowes, IOW
Vernon	Edinburgh
	Portsmouth, Hants
	Brighton, Sussex
Victory	Crystal Palace (WWI)
	Empshott, Hants
	Leamington Spa,
	Warwicks
	Fareham, Hants
	Warsash
	Warnford
	Southsea
	Ventnor, IOW
	Wantage, Berks
	Winchester, Hants
	Idsworth, Hants
	Newbury, Berks
	Porchester, Hants
	Selsey, Sussex
	Yarmouth, IOW
	RN W/T Station,
	Flowerdown
	Woolley Park, Berks

Volcano	Workington, Cumberland		Westcliffe-on-Sea
	Holmrook, Cumbria		Roseneath
Vulture	St Merryn, Cornwall	*Western Isles*	Tobermory, Isle of Mull
Wagtail	Ayr	*Westfield College*	London
Warren	Largs, Ayr Combined Ops	*Wildfire*	Sheerness, Kent
Warrior	Northwood Middlesex		Queenborough Pier,
Wasp	Dover, Kent		Kent
Watchful	Great Yarmouth	*Woolverstone*	Woolverstone, Essex
Wavendon		WRNS HQ	London
Waxwing	Townhill, Fife	*Windsor Castle*	Troopship
Wellesley	Liverpool	*Yeoman*	
Westcliff	Chelmsford, Essex	RNAS Zeals	

INDEX